FRENCH MUSEUM ARCHITECTURE

法国博物馆建筑

策划编辑 **Editorial Consulting**
法国亦西文化 **ICI CONSULTANTS**

总企划 Direction
简嘉玲 Chia-Ling CHIEN

资料收集与整理 Communication / Documentation
尼古拉·布里左 Nicolas BRIZAULT

英文翻译 English Translation
艾莉森·库里佛尔 Alison CULLIFORD

艺术指导 Art Direction
卡琳·德拉梅宗 Karine de La MAISON
维建·诺黑 Wijane NOREE

FRENCH MUSEUM ARCHITECTURE

法国博物馆建筑

ICI CONSULTANTS 法国亦西文化 策划编辑
Alison CULLIFORD 艾莉森·库里佛尔 英文翻译
Catherine CHANG 常文心 中文翻译

辽宁科学技术出版社

INNOVATIONS
IN HISTORIC SPACES
在旧空间里创新

DIALOGUES
BETWEEN OLD AND NEW
新旧对话

BRAND NEW EXPRESSIONS
全新的发挥

Würth Museum: Architect Jacques & Clément Vergély / Sculptor Bernard Venet / photgrapher Erick Saillet

FUTURE PROJECTS
未来项目

INNOVATIONS
IN HISTORIC SPACES

在旧空间里创新

CHAMPOLLION MUSEUM – SCRIPTS OF THE WORLD

商博良博物馆——世界的手稿

建筑师 Architect: Moatti-Rivière
地点 Location: Figeac
完工日期 Completion Date: 2007
摄影师 Photographer: Luc Boegly (pp.8-9), Matthieu Deville (pp.10-13)

The Champollion Museum is housed in four Medieval buildings in the protected sector of Figeac's town centre, one being the birthplace of Champollion, the decipherer of Egyptian hieroglyphics. On its main façade, the arcades partly date back to the 12th century while the upper floors are 18th-century. Covering 1,400m², the museum is composed of a main space displaying the permanent collections, with a temporary exhibition room and an educational workshop situated in independent buildings nearby.

In Moatti et Rivière's renovation, the buildings have been totally restructured to free the museography of all constraints and to open up large visual perspectives. The beams of the new concrete floors are hung from the old façade using metallic struts. These also compose the structure of the loggias and the sun terraces, whose

metallic trellis floors allow one to walk in the space between the two façades.

The existing outer façade has been renovated and ensures the urban continuity with the houses of the town. The second façade, set back around 1m, is made up of 48 glass panels averaging 3.5m by 1.2m. A 0.5-micron copper leaf cut into 14cm x 14cm panels was laid by hand on a polymer film. Each letter was hand-

cut out of the copper. The layered glass contains the copper polymer film between two sheets of PVB. The layers of glass and the different films are combined using a chemical process. This glass was then assembled with a sheet of 8mm safety glass to form a traditional double glazing. The Champollion Museum's aim was to bring the work of artisans together with an industrial process, and the façade conveys the imagination and beauty of the letters to give the project its identity.

The permanent exhibition takes one through seven distinct rooms over four levels. Each room has a thematic unity conveyed by a single colour – black, red orange, blue, ivory – that links the floor, the ceiling and the painted signage on the glass cabinets and the information panels. It allows one to focus on the artefacts in a setting of peaceful unity.

1. Scripts of the Mediterranean room
2. Exterior façade
3. The "soleilo", a typical feature of the region's architecture
4. Master plan
5. The façade of a thousand letters, a play of shadows and light
6. Ground floor plan
7. Typical floor plan

1. 地中海房间的文字
2. 外立面
3. 该地区建筑的一个典型的特征 — soleilo
4. 总体规划图
5. 刻有一千个文字的立面，一场光与影的演出
6. 一楼平面图
7. 标准层平面图

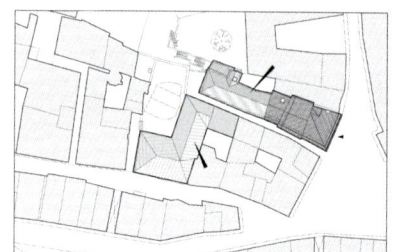

4

5

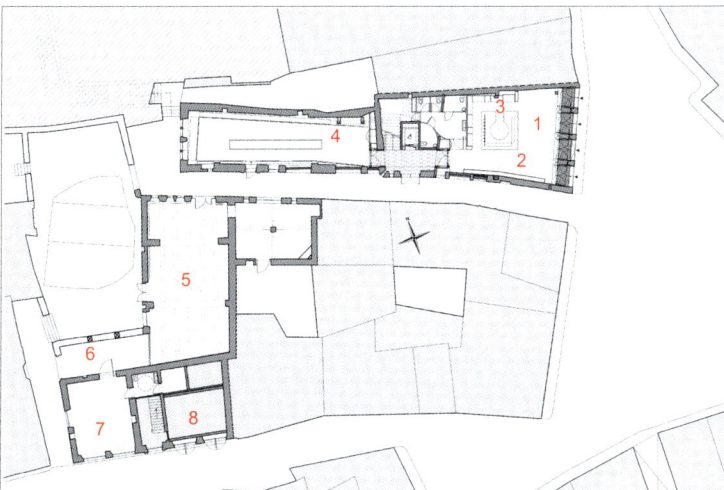

6

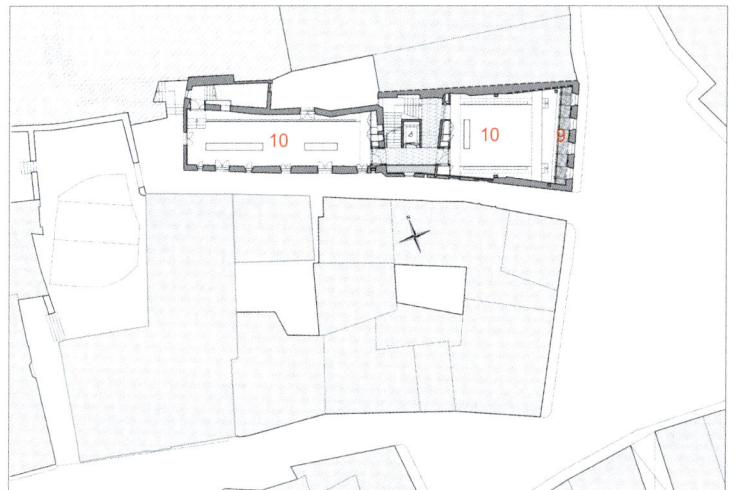

7

1. Waiting/meeting area
2. Reception
3. Ticket desk
4. Champollion room – permanent exhibition
5. Scripts room – temporary exhibitions
6. Gallery
7. Educational workshop
8. Béteille Café
9. Loggia
10. Permanent exhibition room

1. 等待区／会议区
2. 前台
3. 售票窗口
4. 商博良室－永久展览
5. 手稿室－临时展览
6. 画廊
7. 教学工作室
8. Béteille 咖啡
9. 阳台
10. 常设展览室

8

9

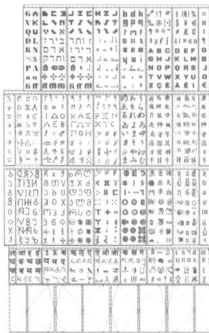

10

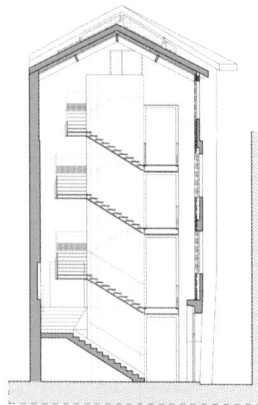

11

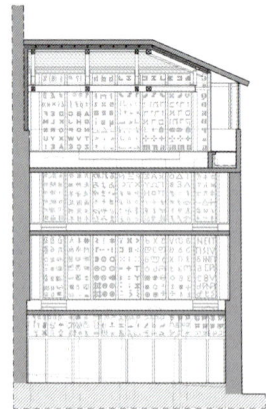

12

8. Scripts of the Mediterranean room
9&13. The palimpsest of the two façades
10. Façade of a thousand letters
11-12. Cross sections

8. 地中海室的文字
9、13. 两面墙呈现的重写本效果
10. 刻有一千个文字的墙体
11、12. 横截面

商博良博物馆分设在菲雅克镇中心保护区的四座中世纪建筑内，其中一座是埃及象形文字学家——商博良的出生地。建筑正面的拱形游廊的历史可以追溯到12世纪，而上面的楼层也可追溯到18世纪。总面积1,400平方米的博物馆有永久性藏品的主展示区、临时展览厅和教学工作室（设在附近的一座独立建筑内）组成。

经过Moatti et Rivière的翻修，建筑释放了所有灵感，打开了视觉透视。新混凝土楼层的横梁通过金属框架从旧外立面上吊下。这些横梁还组成了凉廊和阳台的结构；人们可以通过阳台的金属网格地面在两层外立面之间的空间行走。

原有的外层外墙得到了整修，保证了建筑与城镇住宅的一致性。第二层外墙向后撤约1米，由48块3.5米x1.2米的玻璃板构成。0.5微米厚的铜箔被切割成14厘米x14厘米的面板，通过手工覆盖在高分子膜上。每个文字都由铜箔手工剪切而来。分层玻璃在两层PVB树脂之间加入了铜箔膜。玻璃和不同的薄膜通过化学流程结合在一起。随后，玻璃与8毫米的安全玻璃装配在一起，形成了传统的双层玻璃。商博良博物馆的目标是将工匠的工作通过工业流程结合起来，建筑的外立面传递出文字的形象和美感，树立了项目的形象。

永久展览区覆盖了四层楼，共有七个独立展厅。每个展览厅都有其特有主题色彩——黑色、红橙、蓝色、象牙色，这些标志性色彩连接了地面、天花板以及玻璃陈列柜和信息板上的彩色引导标示。色彩让人们在宁静的整体环境中专注于藏品。

COMIC BOOK MUSEUM
漫画书博物馆

建筑师 Architect: Bodin & Associés
地点 Location: Angoulême
完工日期 Completion Date: 2009
摄影师 Photographer: Enrico Bartolucci
(except n°2), Magelis (n°2)

The Comic Book Museum in Angoulême opened its doors in June 2009, in a series of old wine warehouses beside the Charente. Entirely renovated and enlarged with a new space of 5,000m², it hosts the prestigious permanent collection for which Angoulême is known around the world. One of the major languages of popular art, in the most profound sense of the term, needed to be exhibited, and this called for a museum of contemporary art, no less.

A vast space, clearly lit but respectful of the works, welcomes the visitor at street level to wander among the collections. Its curved route follows the history of the comic book and its major exponents. Through these sinuous presentation areas, the comic book takes on life, presence and sense. Exhibition cabinets at a height that is accessible for all visitors show books, plates and documents marking the great moments and major trends through which this art made its mark on contemporary culture.

Installing itself in the new spaces, the Comic Book Museum has reorganised the whole

of its permanent collections over a space of around 1,300m². With a sober and elegant design that shows the original works to their best effect, the new itinerary is organised in four sections: the first part is devoted to the history of the comic book in the French-speaking world, America and Japan; next "the workshop" is devoted to the different techniques and stages in the creation of a comic strip; "the salon" presents the aesthetic of the comic book; and finally "the gallery" presents thematic exhibitions linked to current events.

昂古莱姆漫画书博物馆选址在查伦太河旁古老的葡萄酒仓库中，于2009年6月正式对外开放。经过全面翻修并增添了5,000平方米的空间，博物馆内珍藏着让昂古莱姆闻名世界的永久性藏品。作为流行艺术的主要类别之一，漫画作品需要通过一个现代艺术博物馆来向世人进行展示，而漫画书博物馆则正好满足了这一需求。

宽敞明亮的空间吸引着参观者漫步于博物馆的藏品之中。蜿蜒的走道沿着漫画书的历史和经典漫画展示缓缓展开。这些迂回的展览区让漫画书变得鲜活生动起来。展示柜的高度方便所有参观者欣赏标志着伟大时刻和主流趋势的书籍、图版和文档，展示了漫画是如何在现代文化中逐步占有一席之地的。

全新的选址让漫画书博物馆在1,300平方米的空间中对全部藏品进行了重新布局。清晰而优雅的设计展示出了原创作品的最佳效果，整个参观路线分为四个部分：第一部分展示法语世界、美国和日本的漫画书历史；"工坊"展示漫画制作的各种技术和阶段；"沙龙"呈现了漫画书的美学价值；"画廊"则主要进行实时的主题展览。

1. Masters of Drawing area – reading point
2. Façade of the museum housed in the 19th-century wine warehouses
3. Perspective view of entrance hall
4. Longitudinal section
5. Entrance hall – staircase leading to the museum

1. 绘画大师主题区的阅读空间
2. 安置于19世纪酒仓建筑内的博物馆立面
3. 入口大厅透视图
4. 纵向剖面图
5. 入口大厅实景 – 通往展览厅的阶梯

administration
administration
administración

↑

auditorium
auditorium
auditorium

5

6

7

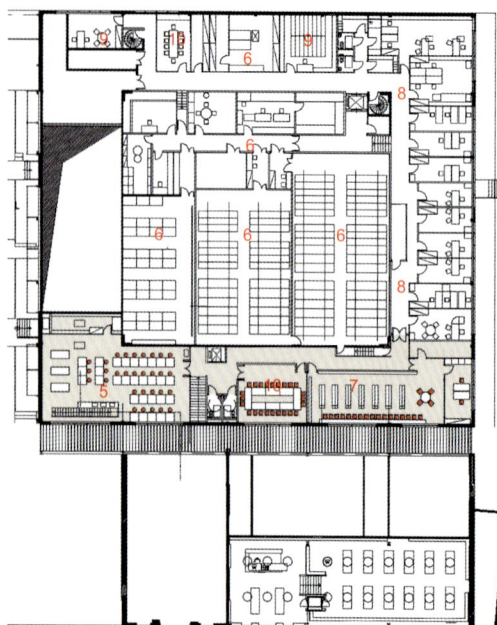

8

6. Perspective view of bookshop
7. Ground floor plan
8. 1st floor plan
9. Bookshop

6. 书店透视图
7. 一楼平面图
8. 二楼平面图
9. 书店

1. Entrance hall
2. Bookshop
3. Permanent exhibition space
4. Temporary exhibition space
5. Mediation service for the public
6. Heritage centre
7. Centre for documentary research
8. Offices
9. Surveillance, technical, upkeep and maintenance
10. Common areas

1. 入口大厅
2. 书店
3. 永久性展览空间
4. 临时性展览空间
5. 多媒体公众服务空间
6. 遗产中心
7. 资料中心
8. 办公室
9. 监视、技术、维修与保养空间
10. 共用空间

12

10. Reading point
11. Comic Books and Society area – daily life reading point
12-13. Comic Book News area – manga reading point

10. 阅读空间
11. 漫画与社会主题区 – 日常生活阅读空间
12、13. 时事主题区 – 日本漫画阅读空间

13

14

15

16

17

14. History area: plastic art – multimedia point and display case
15. History area: display case of "toys"
16-18. Museum reserves

14. 历史主题区：造型艺术－多媒体空间与展示柜
15. 历史主题区："玩具"展示柜
16-18. 馆藏存放空间

18

FABRE MUSEUM
法布尔博物馆

建筑师 Architect: Brochet Lajus Pueyo
地点 Location: Montpellier
完工日期 Completion Date: 2007
摄影师 Photographer: Hervé Abbadie

The renovation of the Fabre Museum aimed to interpret an ambitious project in spatial terms while respecting the "spirit of the place". Comprised of an urban block that brings together three large buildings from different eras, which exhibit different architectural techniques, the project aims to link them together while revealing the uniqueness and the quality of each of its parts.

This main objective was tackled through simple principles. Each of the museum's parts preserves its integrity, is shown in its best light and becomes the basis for a museographical and architectural path through the museum. Each path is clearly identifiable from the entrance hall, imposing a unity of time and unity of place. The courtyards stand out as the strong points of the approach: the "Soulages courtyard", open to the city, together with the interior courtyards of Bourdon, Cabanel, Bazille and Vien, form the high points and the breathing spaces of the composition. The placing of the welcome

hall was crucial, given the number of different floors that it serves. This project establishes it naturally under the Bazille courtyard, the compositional centre of the Jesuit college, on the same level as the Soulages courtyard. It forms the fulcrum of the museum, from which an understanding of the different paths through the museum and of its general organisation extends. The success of the project also comes from the orchestration of the different sequences, which is why great attention was paid to the treatment of the transition

spaces and breathing spaces that are the courtyards.

The "light wing", a real glass showcase, enclosed by the Bourdon courtyard, houses the works the artist Pierre Soulages recently donated to the museum.

The north façade of the Soulages wing is composed of scales of textured glass, lined on the interior by panels of transparent glass. With this "tracing paper" effect the architects have made a contemporary statement in this composite historic building in the heart of the city.

1. The restored grand gallery
2. The Soulages wing (on the right) faces a historic building
3. The Soulages wing – a double glass wall flooded with light serves as a façade to the extension
4. Stone façade facing the street
5. Longitudinal section of the hall and the temporary exhibitions beneath the old courtyards
6. Interior courtyard of the library

1. 翻修的豪华画廊
2. 苏拉基斯翼楼（右侧）面对一座历史悠久的建筑
3. 苏拉基斯翼楼——透光的双层玻璃作为扩建空间的外墙
4. 临街的石材立面
5. 大堂和旧庭院下方的临时展示区纵向剖面图
6. 图书馆的内庭院

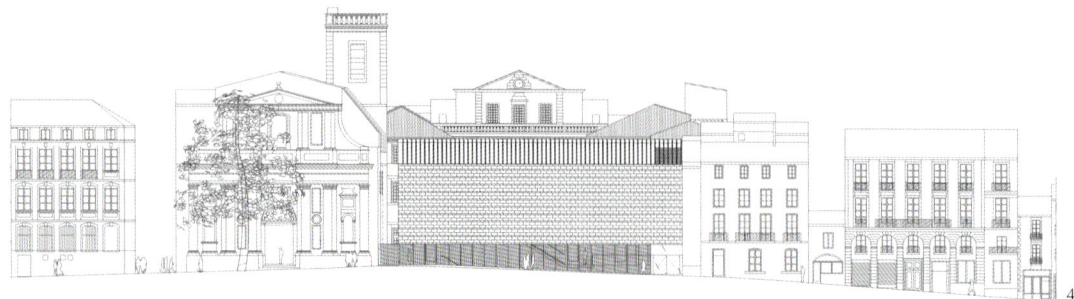

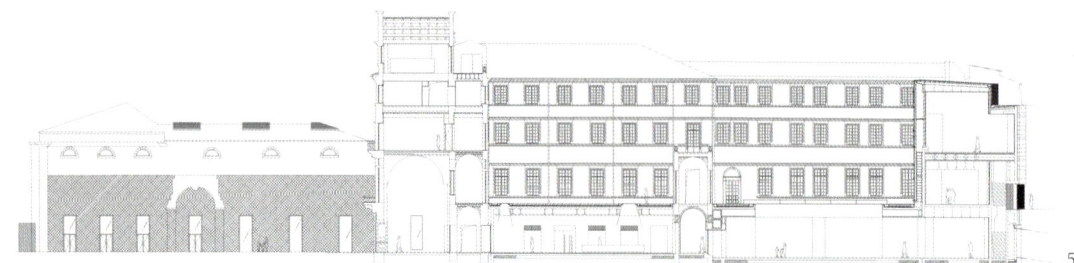

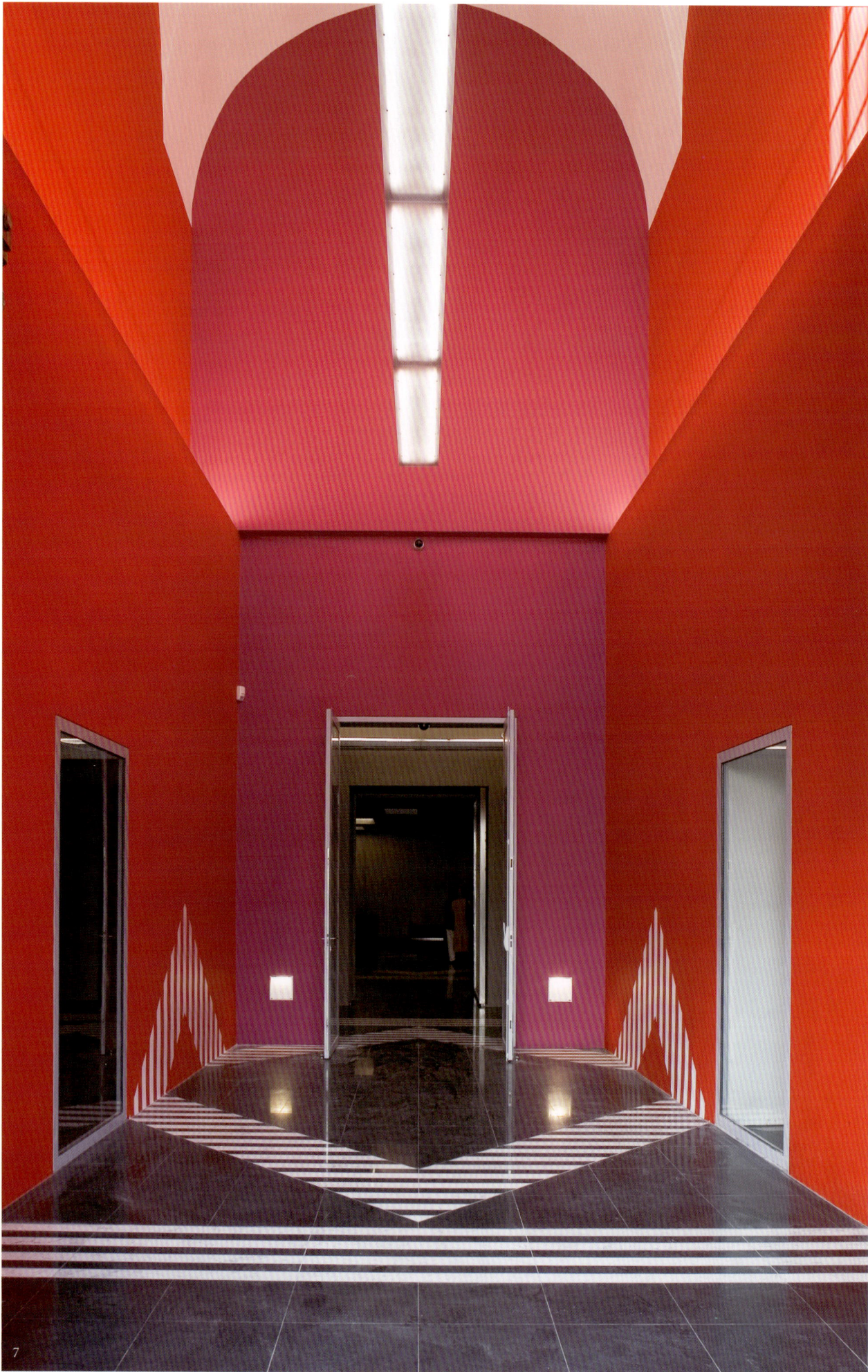

8

7. The entrance hall opens onto a room worked in colour by Daniel Buren
8. The entrance hall in concrete poured in situ slides under a courtyard of the old museum
9. 2nd floor plan
10. 1st floor plan
11. Ground floor plan

7. 入口大厅通往丹尼尔布伦以色彩装饰的厅室
8. 位于旧博物馆庭院下方的水泥墙面入口大堂
9. 三楼平面图
10. 二楼平面图
11. 一楼平面图

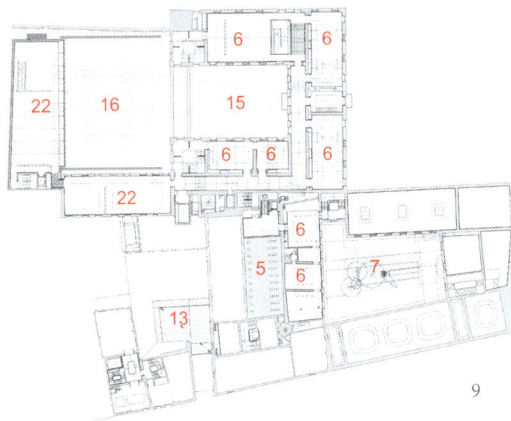

9

10

1. Soulages courtyard (Entrance courtyard)
2. Library
3. Bookshop
4. Entrance hall
5. Sculpture courtyard
6. Exhibition room
7. Vien courtyard
8. Auditorium
9. Restaurant
10. Educational workshops
11. Temporary exhibitions
12. Workshops
13. Cabanel courtyard
14. Permanent collections
15. Bazille courtyard
16. Bourdon courtyard
17. Contemporary gallery
18. Griffins gallery
19. Houdon gallery (old library)
20. Columns gallery
21. Administration & workshop
22. Soulages collection

1. 苏拉基斯庭院（即入口庭院）
2. 图书馆
3. 书店
4. 入口大堂
5. 雕塑庭院
6. 展览室
7. 维恩庭院
8. 礼堂
9. 餐厅
10. 教学工作室
11. 临时展览区
12. 工作室
13. 卡巴内尔庭院
14. 常设收藏区
15. 巴吉尔庭院
16. 博尔顿庭院
17. 现代画廊
18. 格里芬画廊
19. 乌东画廊（旧图书馆）
20. 列柱画廊
21. 行政空间及工作室
22. 苏拉斯收藏区

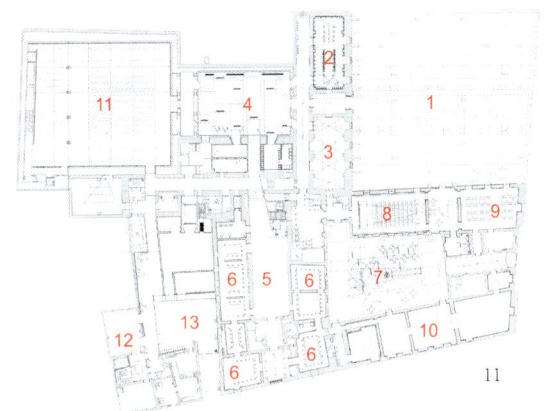

11

12

14

法布尔博物馆翻修工作的宗旨是在尊重"空间的精神"的前提下,从空间上重新诠释这个雄心壮志的项目。项目将三座不同年代的大楼(分别展示不同的建筑艺术)连接起来,旨在展示各个部分的独特价值和品质。

项目的主要目标通过简单的原理得以实现。博物馆的各个部分保留了自身的完整性,在展览路线上体现了博物馆设计与建筑设计的完美结合。每条路线都在门厅清晰可辨,呈现了时间和空间的完整性。庭院在博物馆设计中十分突出:"苏拉基斯庭院"面向城市开放,与博尔顿、卡巴内尔、巴吉尔和维恩几个室内庭院一起形成了博物馆中的休憩空间。由于楼层众多,迎宾大厅的设计

至关重要。项目巧妙地将它设置在巴吉尔庭院——耶稣学院的中心——之下,与苏拉基斯庭院在同一层。它成为了博物馆的支点,人们可以在此辨认出不同路线和整体组织结构。项目的成功还来源于不同序列空间的和谐结合,建筑师在过渡空间和休憩空间(庭院)的处理上倍加留意。

"光之翼"———一个摆放在博尔顿庭院中玻璃展示柜内收藏着艺术家皮埃尔·苏拉基斯新近捐赠给博物馆的作品。苏拉基斯翼楼的外立面由不同尺寸的纹理玻璃组成,内衬着透明玻璃板。这种"描图纸"的效果让建筑师在市中心的历史建筑中成功进行了现代表达。

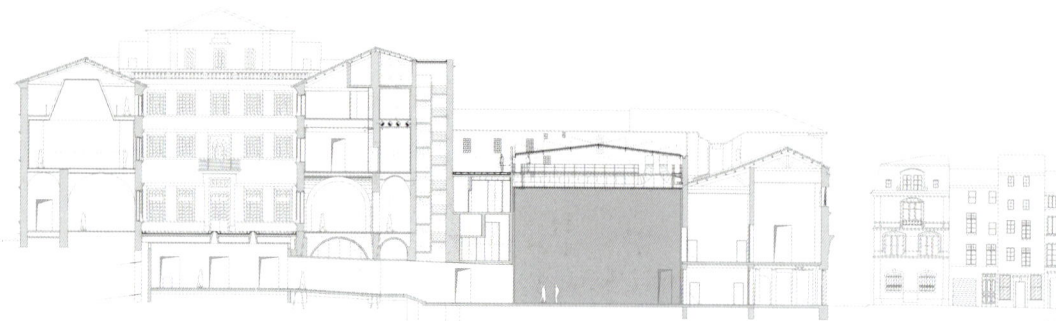

15

16

17

12. The restored Griffins room
13. The Griffins gallery and its restored interior decoration
14. The restored old library
15. Transverse section of the reclaimed and covered courtyards
16. Vaulted galleries of the old convent
17. The restored beams of the galleries
18. Axonometric drawing
19. An old courtyard covered in glass for a sculpture room
20. Simon Hantaï in a concrete alcove
21. The Soulages rooms and natural light distilled by a double skin

12. 翻修的格里芬室
13. 格里芬画廊和翻修的室内装饰
14. 翻修的旧图书馆
15. 被覆盖的庭院截面
16. 古老修道院的拱形画廊
17. 画廊中的翻修梁
18. 轴测图
19. 旧庭院覆盖了玻璃屋顶,成为雕塑室
20. 位于水泥凹室里的西蒙. 汉泰作品
21. 苏拉基斯室和双层玻璃下的自然光效果

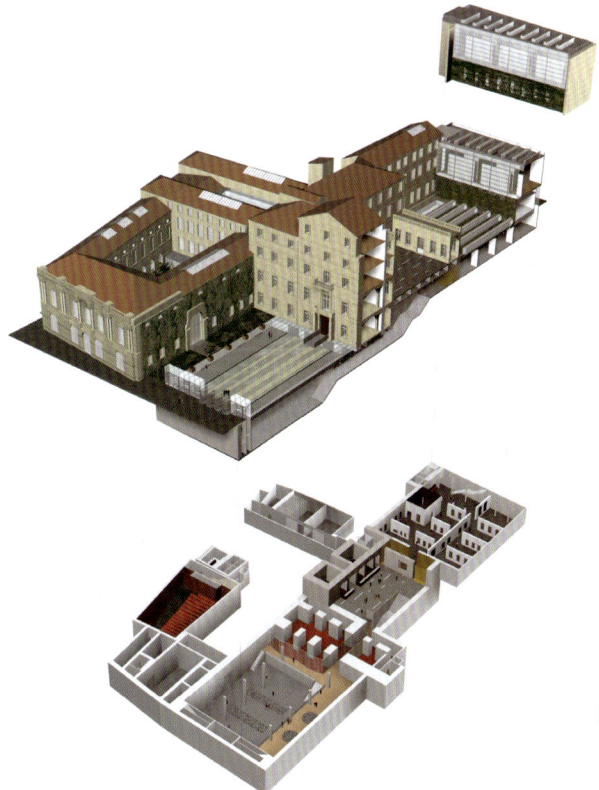

18

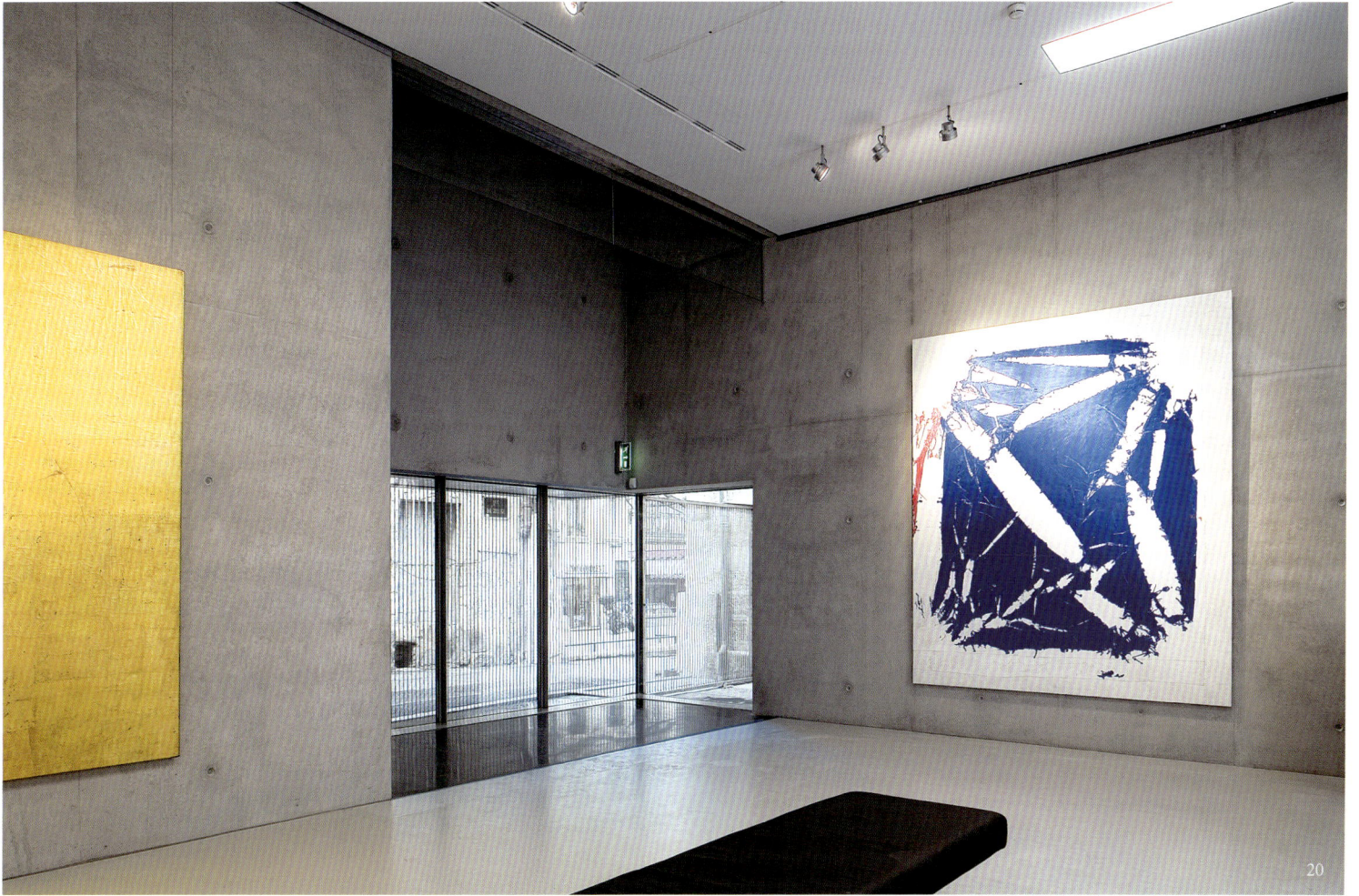

20

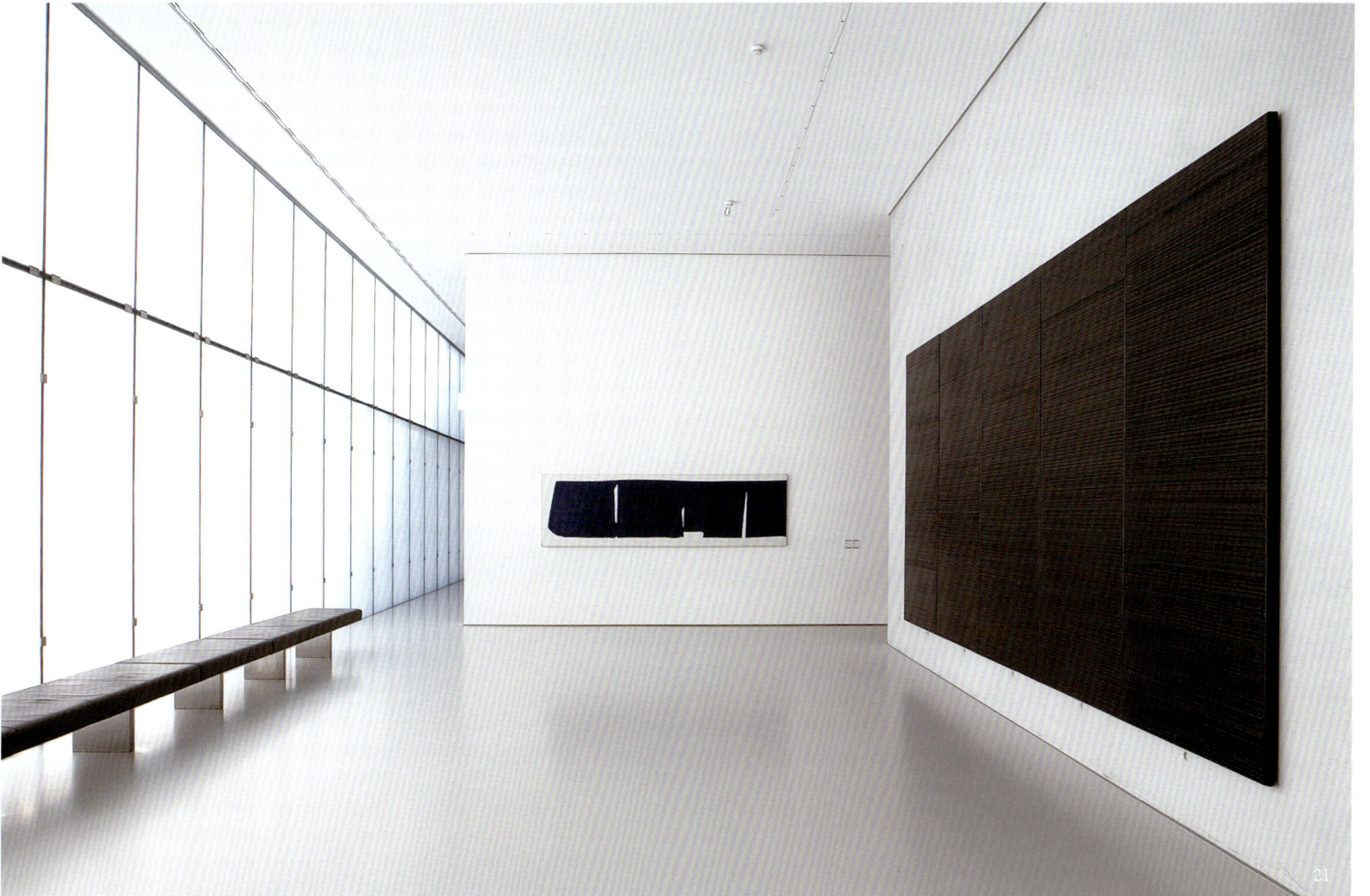

21

HISTORIAL CHARLES DE GAULLE

夏尔·戴高乐纪念馆

建筑师 Architect: Moatti-Rivière
地点 Location: Paris
完工日期 Completion Date: 2008
摄影师 Photographer: Hervé Abbadie

The Historial Charles de Gaulle is situated in the Army Museum, under the Valeur courtyard of the Hôtel des Invalides in Paris. Here, a double challenge is inherent in inserting modernity into a piece of heritage: historical, in terms of communicating the work of Charles de Gaulle; and architectural, with the creation of a contemporary project at the Invalides. The digital architecture of the 21st century has been inserted into a 17th century stone building. Using architecture to create scenes and atmosphere, conveying knowledge and emotion, the Historial is a kind of "audiovisual architecture", where the light of images reveals the material. Interactivity, multimedia and art installations are used to bring archive images to life, as a present-day witness to historical events.

This place of memory is intended to link past and present. Of all the educational tools available to us – books, CD-roms, internet – only museums allow people to immerse themselves physically in the subject. Visitors will compose their own path

of knowledge, whether they are neophytes, curious, passionate about the subject or specialists. Here, space structures time, here an architectural itinerary merges with immersion in the images, monochrome materials merge with multimedia, and historical figures merge with the spectators of history.

The Historial is structured into three sections in successive strata, which complement each other. They represent three ambiances, three experiences for

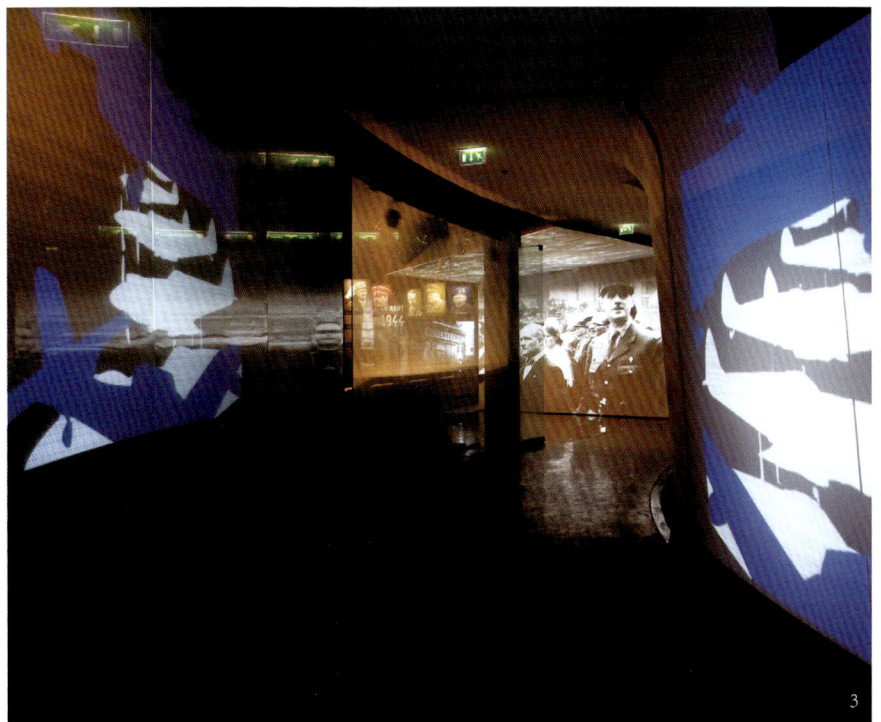

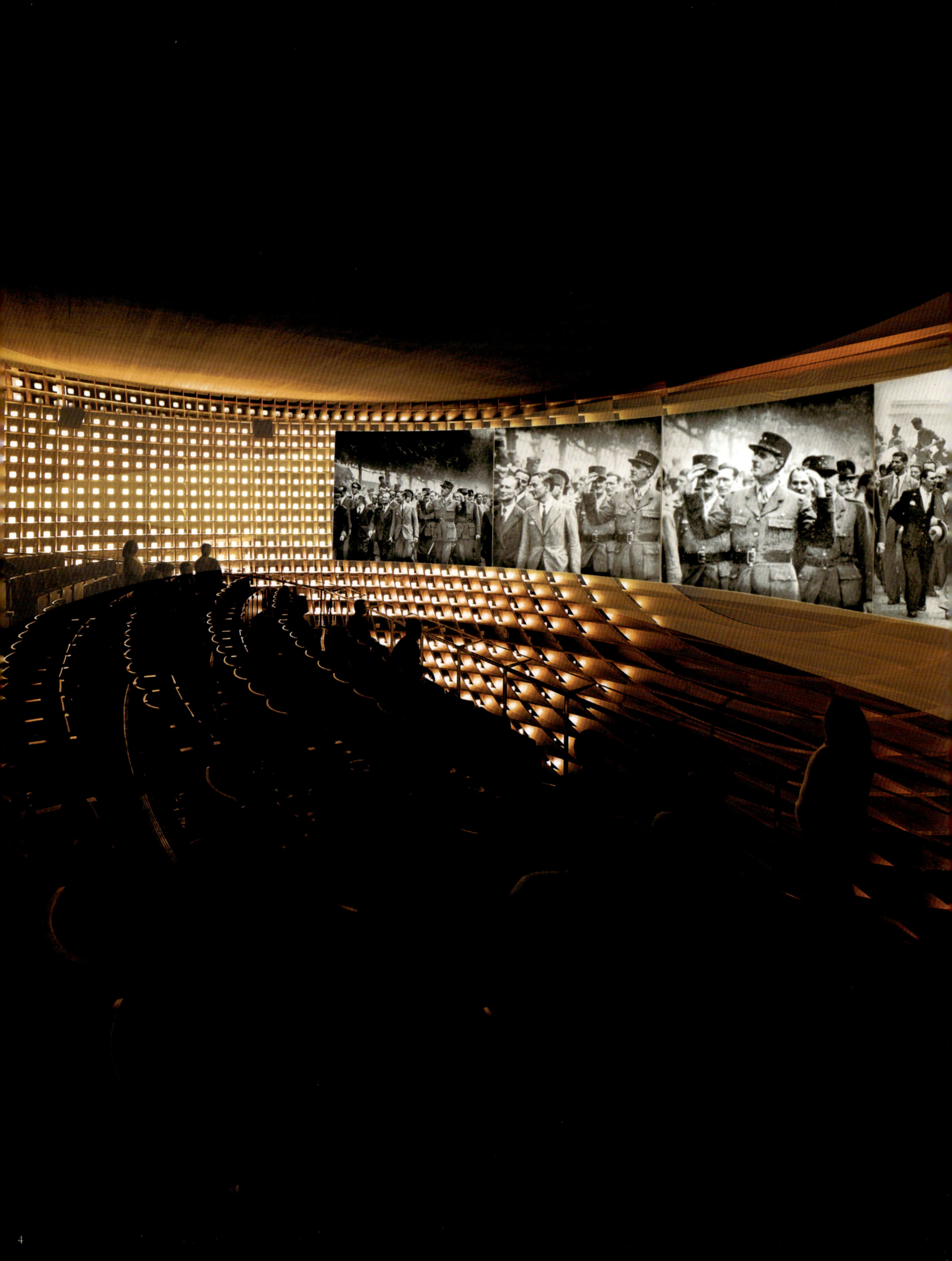

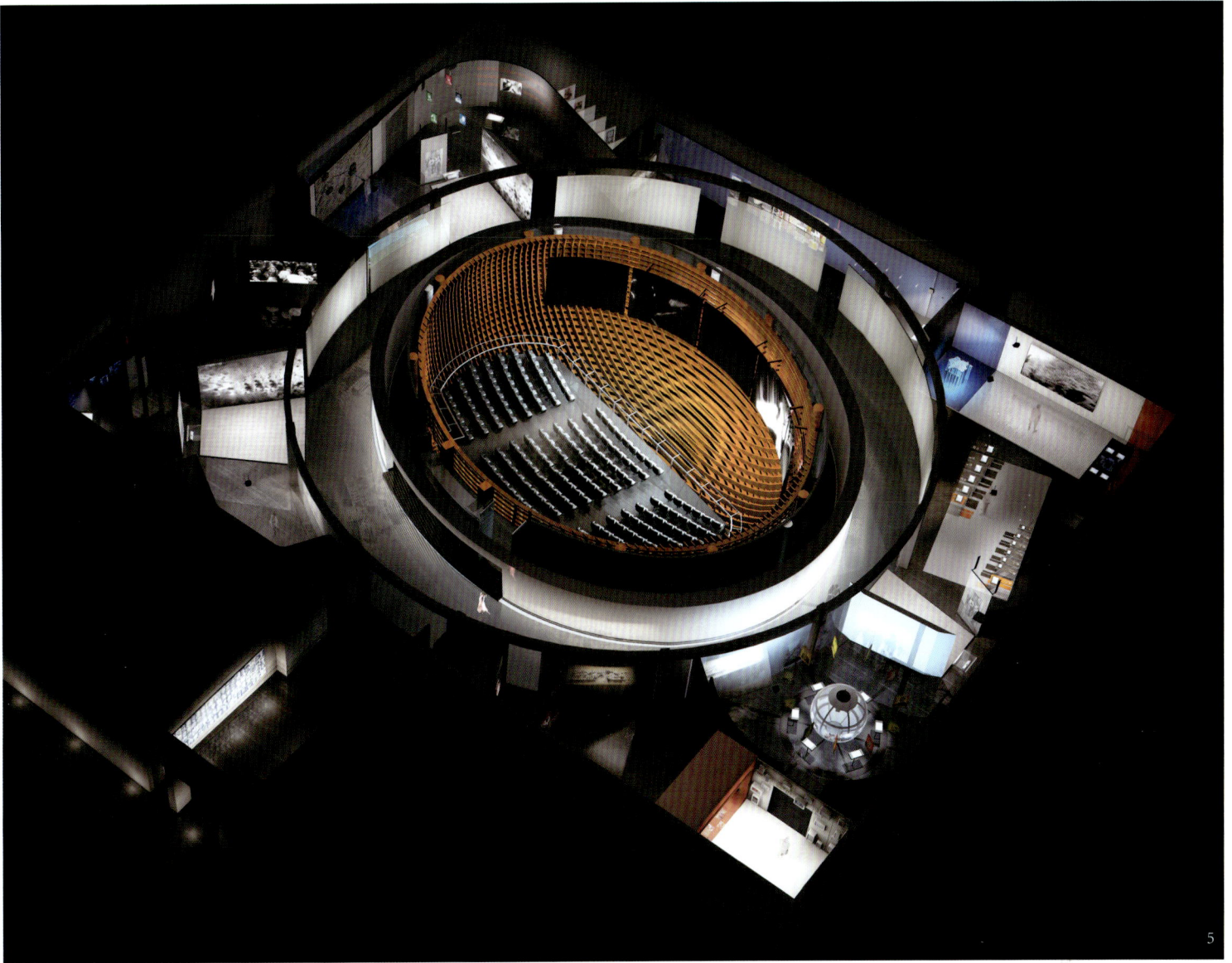

the visitor and three types of access to knowledge. First of all an inverted dome, the multi-screen room where the visitor receives information via a documentary film; next the ring of history, an artistic intervention that delivers us the symbolic images of the 20th century; and finally the three heritage doors, through which the visitor passes from emotion to understanding.

1. The entrance and the corridor leading to the alcoves
2. The ring of history, a junction between the multi-screen room and the alcoves
3. A curved glass screen, crossed by the image, both halts it and remains transparent
4. The multi-screen room
5. General axonometric drawing
6-7. Longitudinal section

1. 直通凹室的入口和走廊
2. 历史环形墙，多屏室和凹室之间的交界处
3. 配以图片的弧形玻璃屏幕，静止且透明
4. 多屏幕室
5. 轴测图
6、7. 纵剖面图

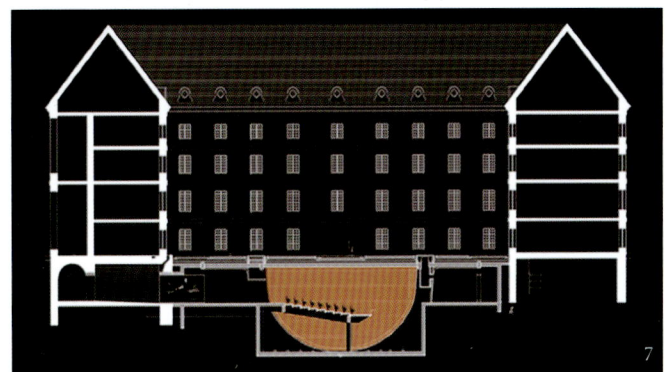

8. In a red space between rooms, the events of May'68
9. The 1960s in France
10. Museographic space dedicated to the Fifth Republic

8. 介于展览室之间的一个红色空间，专门纪念1968年5月事件
9. 法国20世纪60年代
10. 纪念第五共和国的空间

夏尔·戴高乐历史纪念馆位于陆军博物馆内，巴黎荣军院庭院的地下。在为历史遗迹引入现代感的过程中，项目面临双重挑战：表现夏尔·戴高乐事迹的历史感和荣军院的现代建筑结构。21世纪的数字建筑被置入了17世纪的石造建筑之中。纪念馆利用建筑来布景和制造氛围，传递知识和情绪，是一座"视听建筑"。建筑内图像的光线展示了材料。互动性、多媒体和艺术装置让档案图像栩栩如生，让历史事件呈现在眼前。

这个记忆之所连接了过去和现在。在所有可用的教学工具——书籍、CD、网络中，只有博物馆能够让人们身临其境。无论他们是对主题好奇而充满热情的初学者，还是专家、参观者都可以构建自己的知识路线。在这里，空间构建了时间，建筑旅程与大量的图片、单色材料和多媒体、历史人物与历史观察者全都融汇在了一起。

纪念馆划分为三个相辅相成的连续部分。它们代表着三种气氛、三种参观体验和三种获取知识的途径。第一个是倒转的圆顶结构，内部的多屏幕室通过纪录片来传递信息；第二个是历史环，向人们展示20世纪的象征图像；最后是三扇继承门，参观者的情绪由此上升到理解。

ORANGERIE MUSEUM
橘园美术馆

建筑师 Architect: Brochet Lajus Pueyo
地点 Location: Paris
完工日期 Completion Date: 2006
摄影师 Photographer: Hervé Abbadie

Guarding the south-west corner of the Tuileries garden, the Orangerie Museum overlooks the River Seine. It is mirrored by the Jeu de Paume Gallery du in the north-west corner, the two buildings standing either side of the garden's main entrance on Place de la Concorde. The renovation of the Orangerie, completed in 2007, aimed to synthesise the history of the building, which originally housed an orange grove. The architects set out to restore some of its initial character, while respecting the general form of the building and its main lines. The symmetry between the north and south terraces – the Orangerie and the Jeu de Paume Gallery – prevented any addition to the existing building so instead additional gallery space was found by building underground. Wishing to restore some of its original spirit, the architects settled on the idea of an envelope of glass on the south side, towards the Seine, and closed, opaque walls on the north side.

The Orangerie first became a museum after the First World War, when it was

specifically designated to house Claude Monet's large-format Waterlily paintings. In the 1960s, the installation of a second floor for the Walter Guillaume collection literally put Monet's Waterlilies in the shade, thus going against the wishes of the artist. The unique character of Monet's work guided the architects towards an exceptional setting, which provides this museum's main attraction for the public. The work is once again exhibited in natural light, as Claude Monet had originally recommended.

1. The north façade with windows onto the garden
2. Skylight along the north façade
3. The administration offices in a concrete block suspended in the existing volume
4. Entrance hall, north façade with windows onto the Tuileries garden
5. South façade, Seine side. The volume initially planned in wood was built in concrete.

1. 面对花园的北面墙及窗户
2. 北面外墙的玻璃顶罩
3. 行政办公室，上方悬吊着大体积混凝土体块
4. 入口大堂，北面是杜乐丽花园
5. 塞纳河边的南立面。原计划使用木材，最后施工时改为使用混凝土

The Waterlilies remain in place, and their setting has been restored. The interpretation of the "oval hall" described by Monet was the keystone to the approach to light in the whole building. The hall forms a light shaft at the heart of the project, linking all the levels of the museum. It contains the teaching elements needed for an understanding of the work exhibited. This staging of the Waterlily paintings, the reinstating of natural light at the heart of the work, gives the Orangerie its place in the "Grand Louvre" project. Forming the foreground of the perspective towards the Louvre from the quays of the Seine, the Orangerie Museum acts as a lighthouse or a beacon.

6. Approach to the Waterlilies via a wooden bridge
7. The Waterlilies room with its natural light restored
8. Ground floor plan

6. 通过木桥前往"睡莲油画"展厅
7. 恢复自然采光的"睡莲油画"展厅
8. 一楼平面图

橘园美术馆占据着杜伊勒里宫的西南角，远眺着塞纳河。它与西北角的网球场美术馆遥相呼应，两座建筑分列协和广场花园入口的两侧。橘园美术馆的翻修完成于2007年，其目的是综合建筑的历史——场地曾是一片橘树林。在尊重建筑整体造型和主线条的基础上，建筑师恢复了一些橘园的特征。南北两侧露台（橘园美术馆和网球场美术馆）的对称阻止了建筑师在原有建筑上增添任何附加空间，因此，他们将附加的展览空间设在了地下。本着复原原始精神的原则，建筑师在朝向塞纳河的南面采用了玻璃外墙，而北面则采用了封闭的墙壁。

橘园在一战之后被指定为专门收藏克劳德·莫奈的大幅睡莲油画的美术馆。20世纪60年代，沃尔特·纪尧姆的

藏品让莫奈的睡莲黯然失色，违背了艺术家的初衷。莫奈作品特色引导着建筑师进行非凡的布景，成为了博物馆的主要吸引力。美术作品重新被展示在自然光之中，正如克劳德·莫奈的最初建议一样。

睡莲仍在原位，它们的布景却经过了修复。莫奈所描绘的"椭圆大厅"是整栋建筑自然照明的基石。大厅在项目中央起到了光井的作用，连接了博物馆的各个楼层。大厅内包含着便于理解展示作品的教学元素。睡莲油画的展示、自然采光中心地位的恢复让橘园美术馆成为了"大罗浮宫"项目的一部分。从塞纳河的港口上看，橘园美术馆列在罗浮宫前，起到了灯塔的作用。

1. Entrance hall
2. Ticket desk
3. "Drawbridge"
4. Vestibule
5. Waterlilies room
6. Skylight over the permanent exhibition gallery

1. 入口大堂
2. 售票处
3. 过桥
4. 前厅
5. 睡莲油画展厅
6. 永久性展廊上方的玻璃顶罩

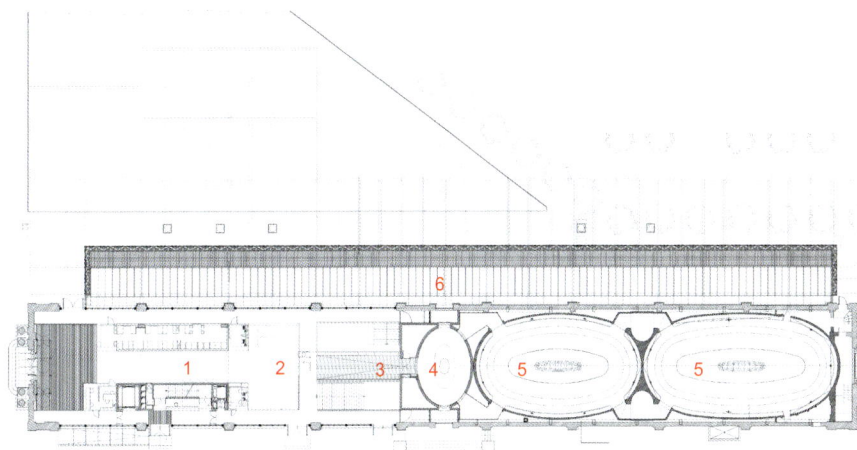

9. The mezzanine bookshop
10. Basement floor plan
11. Longitudinal section of the entrance hall and the Waterlilies room
12. North gallery created beneath the Tuileries garden
13. The Soutine room beneath the garden

9. 阁楼书店
10. 地下室平面图
11. 入口大堂和"睡莲油画"展厅的纵剖面
12. 位于杜乐丽花园下的北廊
13. 花园下方的苏蒂纳室

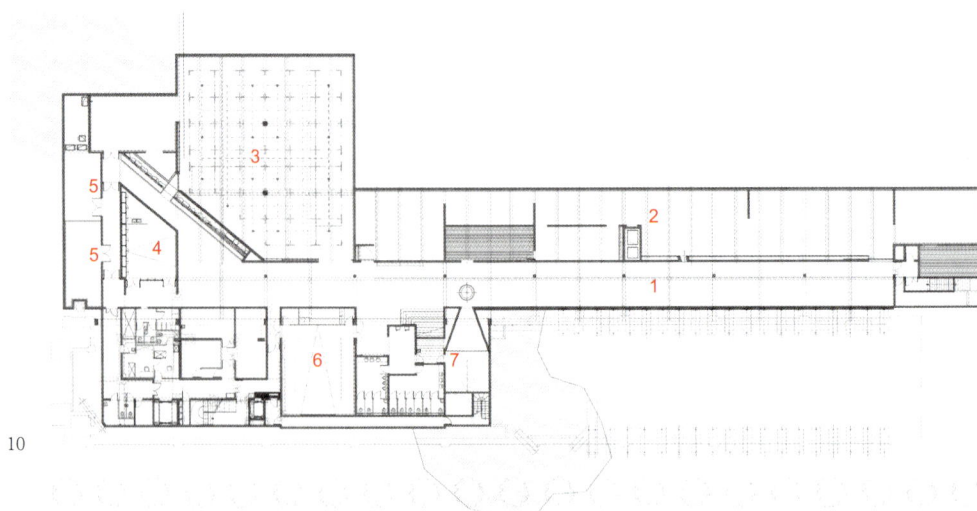

1. Permanent exhibition gallery
2. Permanent exhibition rooms
3. Temporary exhibition rooms
4. Educational room
5. Reserves
6. Auditorium
7. Introduction room

1. 永久性展廊
2. 永久性展览室
3. 临时性展览室
4. 教学室
5. 储藏空间
6. 大讲堂
7. 展览简介室

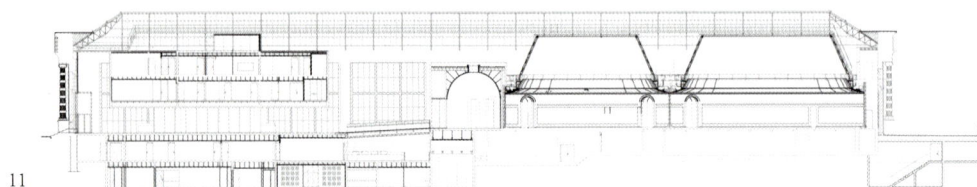

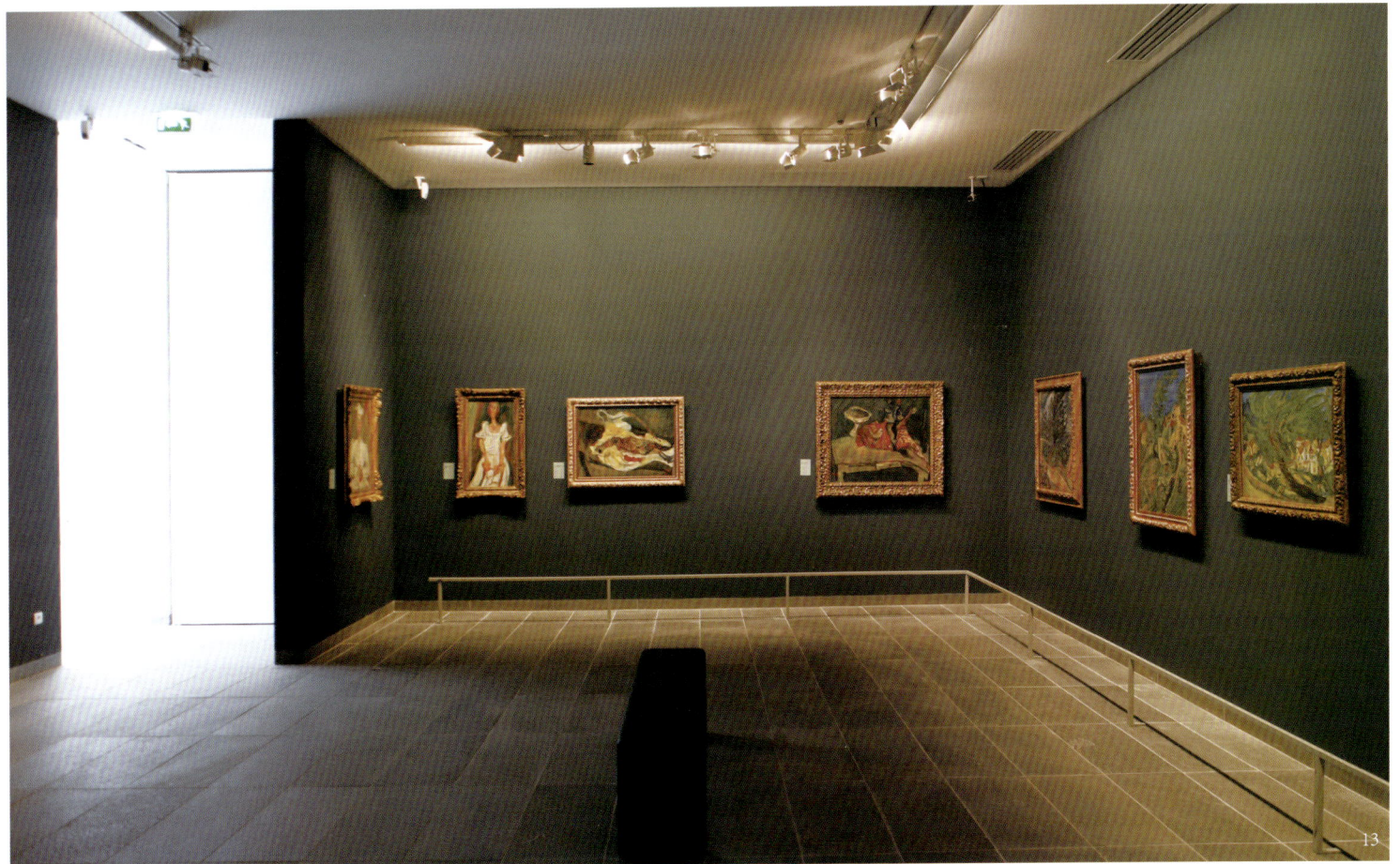

ORSAY MUSEUM
奥赛博物馆

建筑师 Architect: Jean-Paul Philippon
地点 Location: Paris
完工日期 Completion Date: 1986
摄影师 Photographer: Wijane Noree
(pp.50-51), Deidi von Schaewen (pp.52-55)

In 1979 the three architects Renaud Bardon, Pierre Colboc and Jean-Paul Philippon won the competition to transform the former Orsay train station into a museum. Their approach, which grew out of their research paper "Metamorphosis of the Architectural Object", presented to the Ministry of Culture in 1977, was based on three principles:
- the city is a work in constant metamorphosis, which has an effect on every building;
- transforming a building to recycle it in urban life is legitimate;

- the dialectic between the pre-existing architecture and contemporary contributions will generate a new architecture.

Since the unloved station had been earmarked for destruction only a short time before, demonstrating that it could be transformed into a museum was decisive in saving it. The project addressed the enormity of the vast hall and brought the immensity of the whole building down to the scale of the artworks exhibited, it

remodelled the large space under the glass roof of the old station into a succession of rooms and galleries spread over two floors, along an inclined central avenue from which the visitor could easily move from place to place.

A geometrical transformation has been achieved, based on the station's original architect Victor Laloux's very clear composition. Used along its longitudinal axis, the full scale of the grand nave can be appreciated, with the Bellechasse forecourt

and the glass canopy dedicated to welcoming visitors. Overlooking the Seine on its north side and lit by overhead natural light in the old eaves of the station hall, the Impressionists gallery runs between the Amont and Aval pavilions, opposite the Louvre. Natural light infiltrates the building through the station's glass roofs. It is made more atmospheric, finished and assisted by indirect lighting. Vertical planes of ochre Buxy stone reflect the light and add warm tones. They frame Laloux's architecture in surprising ways. Inventing his own itinerary, the visitor simultaneously takes in the old building, the new construction and the artworks exhibited. The station has given the museum its unusual spaces, and the museum has saved the station.

1. A Parisian station transformed into a museum – view of the entrance
2. Seine side façade
3-5. The nave with its sculptures. Natural light infiltrates the building through the station's glass roofs
6. Axonometric drawing of the nave
7-8. Statues in the nave, a central avenue from which the visitor could easily move from place to place

1. 一个巴黎火车站改建为博物馆——入口处
2. 临塞纳河的建筑立面
3–5. 自然光透过火车站的玻璃屋顶射进陈列着雕塑品的中殿
6. 中殿轴测图
7–8. 中殿有如一条雕塑大道，参观者经由这里而通达其他展览空间

1979年，赫诺·塞尚、皮埃尔·科波克和让-保罗·菲里朋三位建筑师获得了设计竞赛的优胜，将前奥赛火车站改造成为一座博物馆。他们的设计方案来自于调研报告《建筑目标变形记》（于1977年上呈文化部），以三项基本原则为基础：

– 城市是不断变形的产物，对每一座建筑都有所影响
– 改造建筑，使其在城市生活中循环是合理的
– 已存在建筑和当代改造将创造出一座全新的建筑结构

车站在改造前不久还被宣布将进行拆除，将它改造成博物馆无疑是拯救了它。项目突出了大厅的宏伟感，并且将建筑的宏大感缩减到展览艺术品的规模。它将旧车站玻璃屋顶下的巨大空间改造成一系列房间和画廊，参观者们可以沿着一条倾斜的中轴便利地穿梭。

项目的几何改造以建筑师维克多·拉卢对车站的清晰布局为基础。沿着原始设计的纵轴，人们可以欣赏到中殿的宏伟，柏歇斯前院和玻璃穹顶都欢迎着宾客。印象派画廊贯穿阿蒙山崖馆和阿瓦尔馆，俯瞰着塞纳河北岸，通过车站屋檐上透过来的自然光照明，正对罗浮宫。自然光透过车站的玻璃屋顶进入建筑，与间接照明一起营造了更和谐的气氛。垂直的褐色泊斯石反射着光线，增添了温暖的色调。它们以惊艳的方式将拉卢的建筑包裹起来。参观者可以制定自己的参观路线，同时欣赏旧建筑、新结构和艺术品。车站赋予了博物馆非同寻常的空间，而博物馆则拯救了车站。

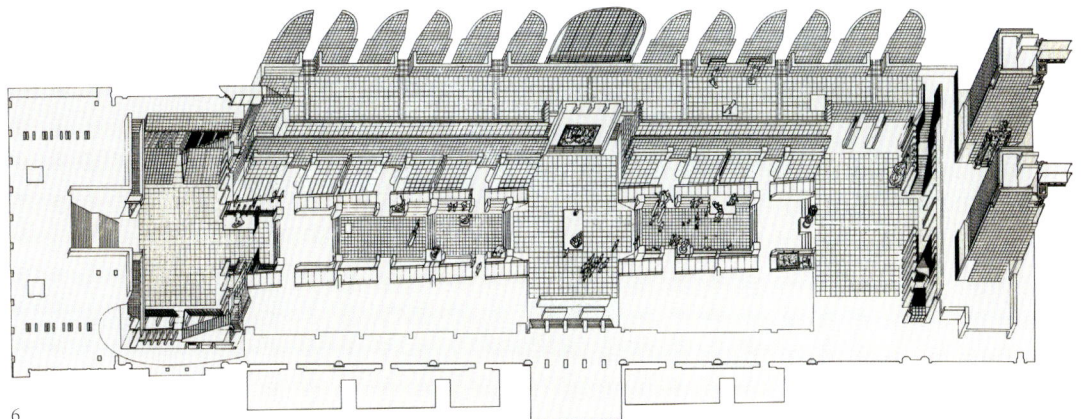

6

AMONT PAVILION
AT THE ORSAY MUSEUM

奥赛博物馆阿蒙馆

建筑师 Architect: Atelier de l'Ile
地点 Location: Paris
完工日期 Completion Date: 2011
摄影师 Photographer: Atelier de l'Ile,
Hervé Abbadie

In 2011, Atelier de l'Ile redesigned the Amont Pavilion to provide a new setting for decorative arts, the Nabis painters and the large format paintings of Gustave Courbet. Drawn in by the red of the wall at the end of the nave, the visitor enters a pared down environment where everything has been designed to show the artworks to their best advantage.

The Pavilion was open to the rest of the museum and linked to it via passages and a footbridge. A simple and legible circuit for visitors has been created through a new approach to vertical circulation and a circular route on each floor. From the ground floor, the way up to the Impressionists gallery at the foot of the tympanum has been made more legible by grouping the staircase and lift with the escalators at the end of the nave, thus reducing the number of pillars and

obstacles in this part of the museum. From the larger surface area of the rooms to the uncluttered and discreet museography, everything works to provide an improved setting for the artworks. The organisation of each floor is now much clearer thanks to a sharp distinction made between the exhibition spaces and the circulation hubs: for the latter, a clear path through the space is strengthened by natural light, while state-of-the-art artificial light is deftly

Décors modernes
1905-1914

5

1. Vertical view down the well of the central staircase
2. Sculpture and light
3. Exterior of the pavilion
4. Staircase and footbridge linking to the nave
5. 2nd floor, the red theme and the light wells
6. Axonometric drawing
7. Position of the Amont Pavilion

1. 中央楼梯垂直视角
2. 雕塑与光线
3. 展馆外观
4. 连接中殿的楼梯和人行桥
5. 三楼的红色主题和采光井
6. 轴测图
7. 阿蒙馆的位置

employed in the exhibition rooms.

The space is brought alive by the use of different scales. The floors dedicated to the different schools of decorative arts have been reworked on the scale of the furniture and objects exhibited. The structure of the rooms and the display supports (plinths, display cases…) are conducive to representing the works "in situ", recreating the intimate atmosphere of a private home. Conversely, to reveal the clock and the exceptional view over the Seine in all their glory, the 5th floor space has been cleared of obstacles. In the same way, Courbet's large format paintings have been given a room that they deserve, with a large circular seat the only furniture. Finally, work on colour has been done, with the red wall serving as a clarion call, a landmark, a means of getting one's bearings, and a backdrop to the vertical circulation, while a dark "taupe" hue has been used for the gallery walls, the better to show the artworks.

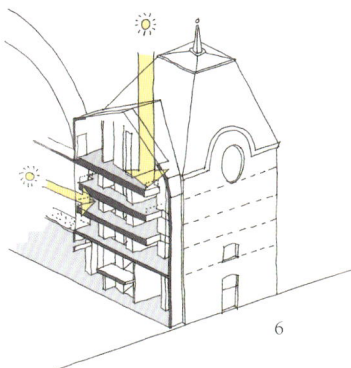

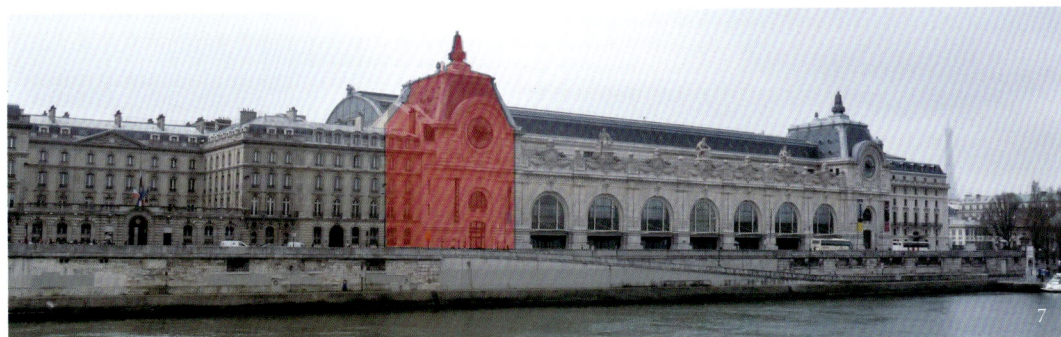

8

2011年，l'Ile工作室重新设计了阿蒙馆，为装饰艺术、那比派画家和古斯塔夫·库尔贝的大幅油画提供了全新的布景。受到中殿尽头的红墙吸引，参观者会进入一个缩小的空间，那里的一切都是为了以最佳方式展出艺术品而设计的。

展馆面向博物馆的其他部分开放，通过通道和人行桥相连。参观者可以通过简单可辨的路线对展馆进行参观，每层楼的路线都呈圆形展开。从一楼通往印象派画廊的路线十分明显，楼梯、电梯和自动扶梯全部设在中殿后部，减少了展览空间内的柱子和障碍数量。从展厅的宏大面积到整齐的博物馆布置，一切都是为了更好地展现艺术品。展览空间和流通中心的鲜明对比让每层楼的组织结构变得异常清晰。流通中心的自然采光凸显了走道的清晰性，而展览厅则采用了先进的人工照明。

不同尺度的运用让空间活跃起来。不同装饰艺术学派的展示楼层都与其展出的家具和物品进行了匹配。空间的结构和展览设施（底座、展示柜……）都有利于现场展示作品，重现私密的家居气氛。于此相反，为了展示时钟和塞纳河的非凡美景，六楼的空间没有丝毫阻碍。同样的，库尔贝的大幅作品被展示在恰当的空间，巨大的圆形座椅是唯一的家具。最后，在色彩选择方面，红色的墙壁起到了召唤作用，以地标的形式吸引了人们的注意力，也为垂直交通提供了背景；画廊的墙壁则采用了深灰褐色，以便更好地展示艺术品。

8. Natural light and a red wall (red line)
9. New vertical circulations
10. Vertical view of the atrium and red walls (red line) on a grand scale

8. 自然光和红色墙（红带）
9. 新的垂直交通空间
10. 中庭和红色墙（红带）的垂直视图

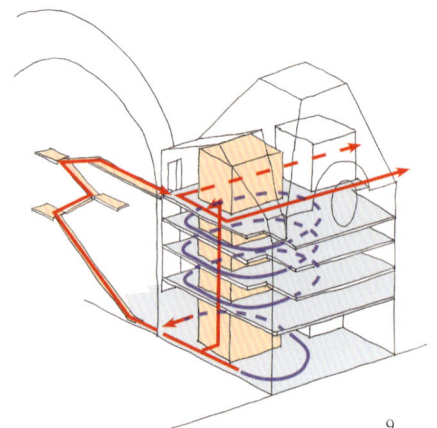

9

11. 2nd floor, the Nabis
12. 4th floor, European furniture
13. 2nd floor, the Nabis, the different arts brought together
14. 3th floor plan, thematic exhibition room
15. Section

11. 第三层的纳比派
12. 第五层的欧式家具
13. 第三层的纳比派，汇集了不同艺术品
14. 四楼平面图，主题展览室
15. 剖面图

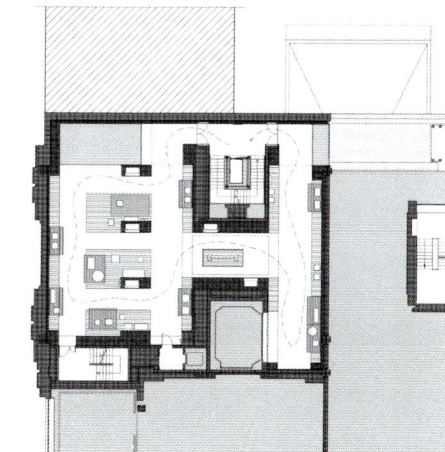

14

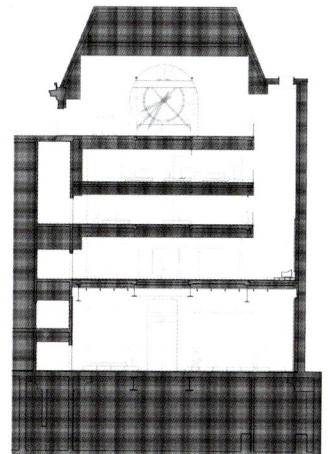

15

16

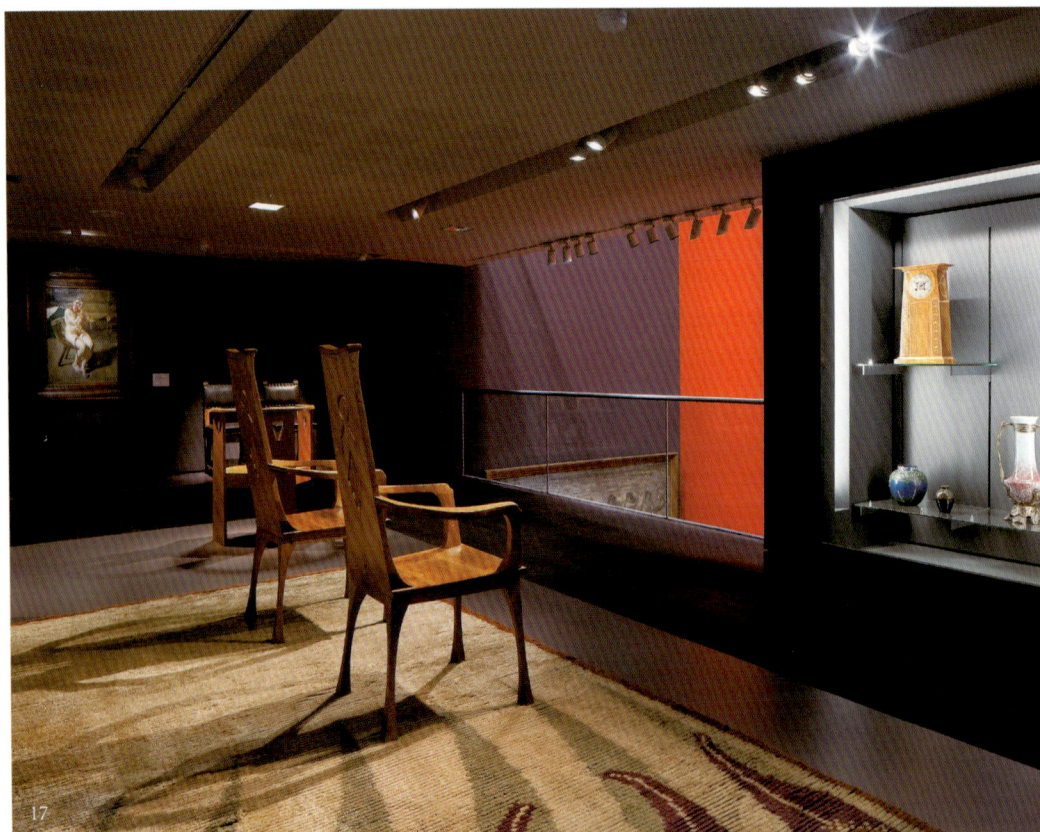

17

16. 4th floor, display cases and European furniture (Austria)
17. Natural light and artificial lighting, on a domestic scale
18. The central staircase and the lift shaft in metallic mesh
19. 5th floor, the clock, the skylight and the atrium
20. 5th floor, the clock and the view of Paris

16. 第五层，陈列柜和欧洲的家具（奥地利）
17. 室内的自然观和人工照明
18. 中央楼梯和金属网内的电梯井
19. 第六层的大钟、天窗和中庭
20. 透过第六层的大钟可眺望巴黎景色

18

19

20

INTERNATIONAL PERFUME MUSEUM

国际香水博物馆

建筑师 Architect: Frédéric Jung
地点 Location: Grasse
完工日期 Completion Date: 2009
摄影师 Photographer: Frédéric Jung
(pp.66-67, 69, 70, 75), Éric Laignel
(pp.68, 71-73, 76-77)

The challenge in designing this museum was to convey the subject of perfume, newly recognised as a form of heritage but nevertheless intangible. This museum does not limit itself to conservation, but sees itself as a tool – an "interpretation centre" housing a collection and offering the means to read it in a contemporary way. The permanent exhibition of the collections is fed by satellites: a lecture room, temporary exhibitions, an osmotheque, an archive, educational workshops for children and adults, etc.

This gave rise to a complex project, spread over multiple buildings, marked by the history of the town and its successive additions. Placed on a hillside, perpendicular to a topographical fault, on the edge of the old city fortifications, the museum had to make good use of this spatial and historical richness. The primary stake in its reorganisation was how to create a relationship across the 3,000m^2 spread over five buildings and seven floors. A major reference point creates a synergy for the many different spaces. It

has been created around the uncovered 14th-century city wall, enshrined at the heart of the block by a "fault" that offers a clear view of the exhibition floors and organises the movement through the museum.

Paradoxically, it is a wall around two metres thick that becomes the point of contact, the link for the new organism. This axis of fortification is revived and protected by a glass "nave" running alongside it and supported by it. The nave appears as the

contemporary face of the museum from Boulevard Fragonard, while not interfering with the lines of the Hôtel Morel-Amic, the master building of the site. The itinerary through the museum is extremely varied. Neither a temple-museum or not forum-museum, it tells a series of "stories" in a living way, in seven very different settings.

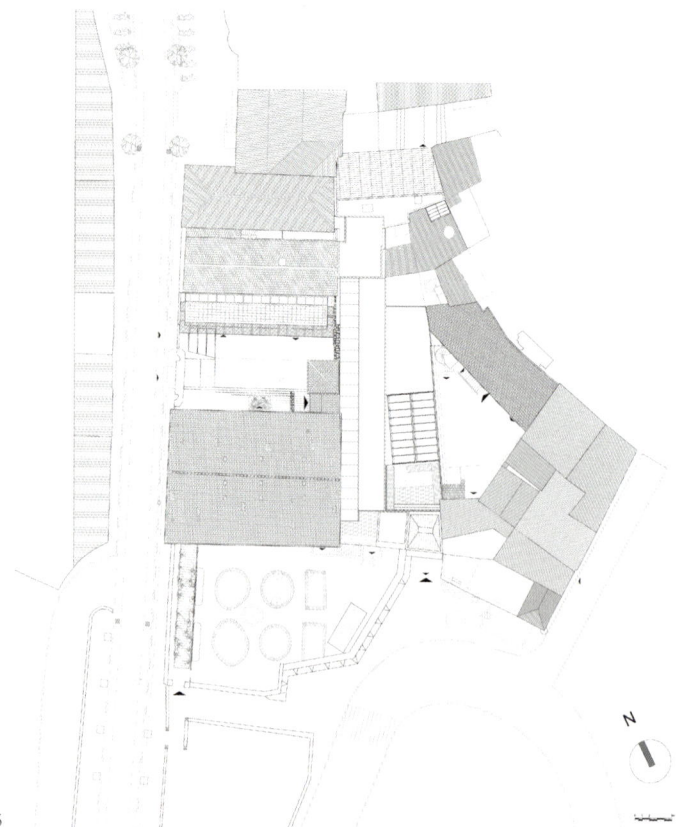

6

1. Façade on Jeu de Ballon boulevard
2. South façade
3. The fault
4. Entrance hall
5. Master plan
6. Entrance to the temporary exhibitions

1. Jeu de Ballon大道上的立面
2. 南立面
3. 断层空间
4. 入口大堂
5. 总体规划图
6. 临时展厅入口

博物馆设计的挑战在于传递香水——这一无形的传承文化——的主题。博物馆并不局限在保存功能，而是视自己为工具，以"解说中心"的身份收藏香水并让人以现代的方式对香水进行解读。除了永久展览区之外，博物馆还配所有讲座厅、临时展览区、香水储藏厅、档案馆、儿童和成年人的教学工坊等。

这个综合的博物馆项目由若干座建筑组成，以城镇的历史和后续叠加的扩建项目为特色。博物馆坐落在山坡的断层上，位于古城的防御工事边缘，其设计必须充分利用空间和历史的丰富价值。博物馆重组的主要问题在于如何在五座楼、七层楼、3,000平方米的空间之间建立联

系。建筑师在14世纪的城墙上建立了一个协同各个空间的参照点，在城墙中央建立了一个断层。人们可以从断层上清晰地看到展览楼层和博物馆组织结构。

这座两米厚的墙成了连接点，将全新的有机体联系在一起。防御工事的中轴重获新生，由一个玻璃"中殿"所保护和支撑。从弗拉戈纳尔大道上看，中殿是博物馆的现代外观。同时，中殿又不会破坏场地上主建筑——莫雷尔酒店的线条。贯穿博物馆的路线变化多端。香水博物馆既不像庙宇，又不像论坛，它在奇怪不同的布景中以活泼的方式讲述着一系列故事。

7. The terrace
8. Longitudinal section (contemporary displays + rooftop conservatory)
9-10. Rooftop conservatory

7. 屋顶露台
8. 纵向剖面图（当代展示区和屋顶温室）
9、10. 屋顶温室

11. The fault
12-14. Industrial machinery in the fault

11. 断层空间
12-14. 设置在断层空间里的工业机械

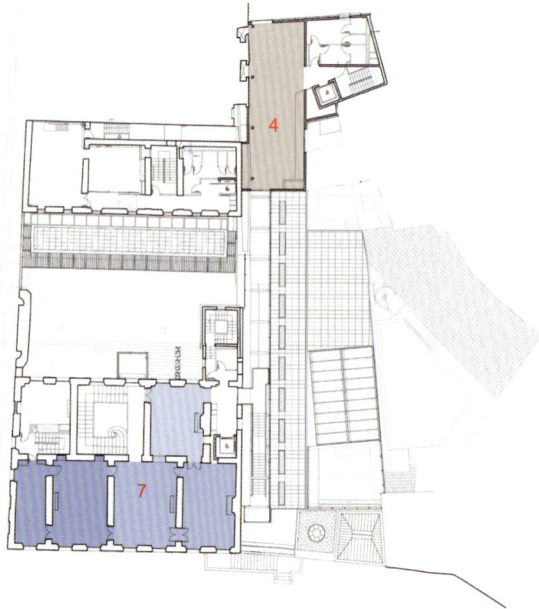

15

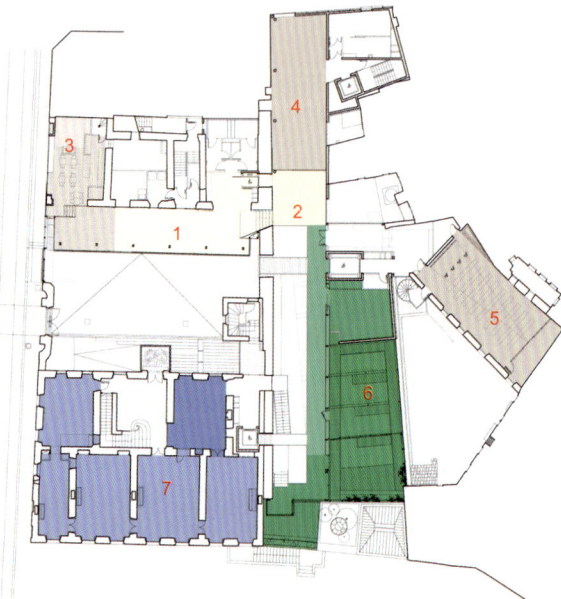

16

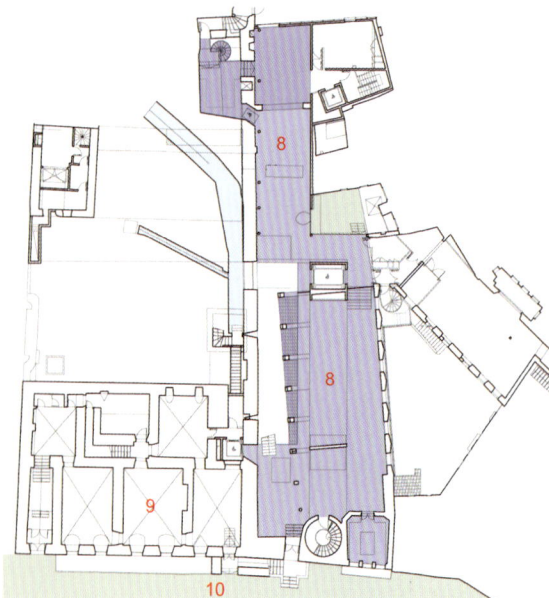

17

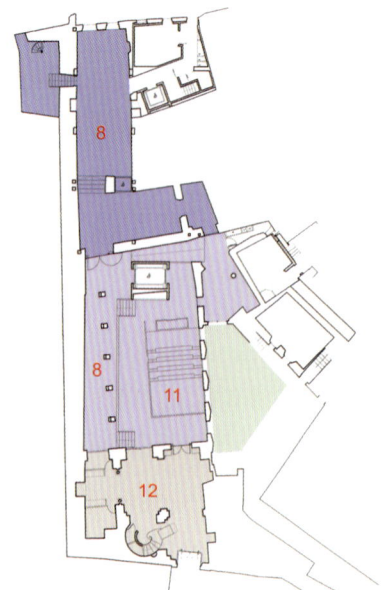

18

19

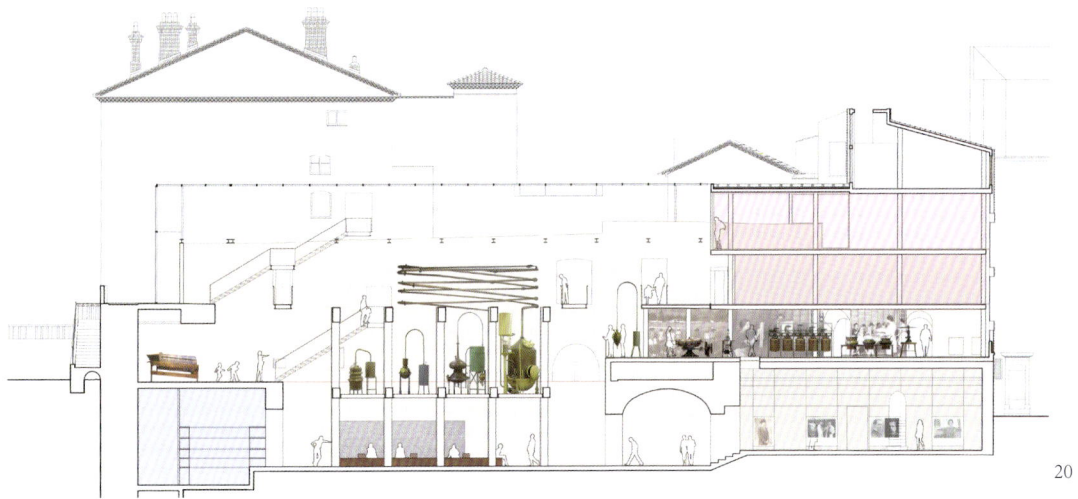

20

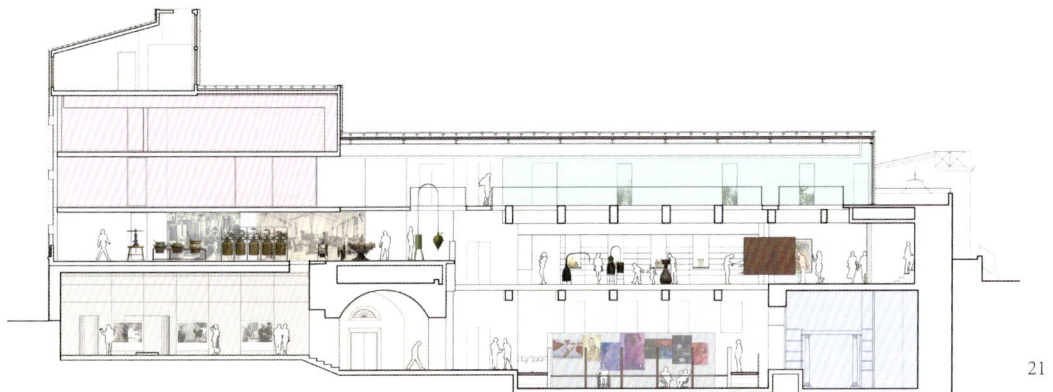

21

22

22. 19th century displays
23. Early 20th century displays
24. 20th century: a perfume for each year

22. 19世纪展示区
23. 20世纪早期展示区
24. 自20世纪以来每一年推出的香水展示

TOMI UNGERER MUSEUM

汤米·温格尔美术馆

建筑师 Architect: ECDM – Emmanuel
Combarel Dominique Marrec
地点 Location: Strasbourg
完工日期 Completion Date: 2007
摄影师 Photographer: Luc Boegly

Situated in the heart the Ville Neuve district, near Place de la République, the Tomi Ungerer Museum and International Centre for Illustration is housed in the Villa Greiner, which was built in 1884 by the architect Samuel Revel. The building's neo-classical style is typical of the late19th-century architecture of this German Imperial neighbourhood and as such had to be respected, whilst adapting it to its new role as a major cultural destination for Strasbourg and its region.

The bourgeois residential building thus changed its status to become a public building, a space to welcome visitors ranging from researchers and academics to the simply curious. The architects endeavoured to take on board the regulatory demands and technical contingencies while giving the building a specific identity, that of a cultural space in an overall architectural context.

It is a contemporary intervention in a 19th-century setting. The integrity of the exterior

of the building has been preserved, the only change being to the windows, which, as an interface, had to be updated to respond to the specifications of the museum, and the installation of a curved walkway across the garden.

Inside, the proportions of the original rooms have been preserved, but white ceilings, floors and walls give them a "white cube" aesthetic suitable for the display of graphic works. Visitors can easily move between the four floors of the museum via

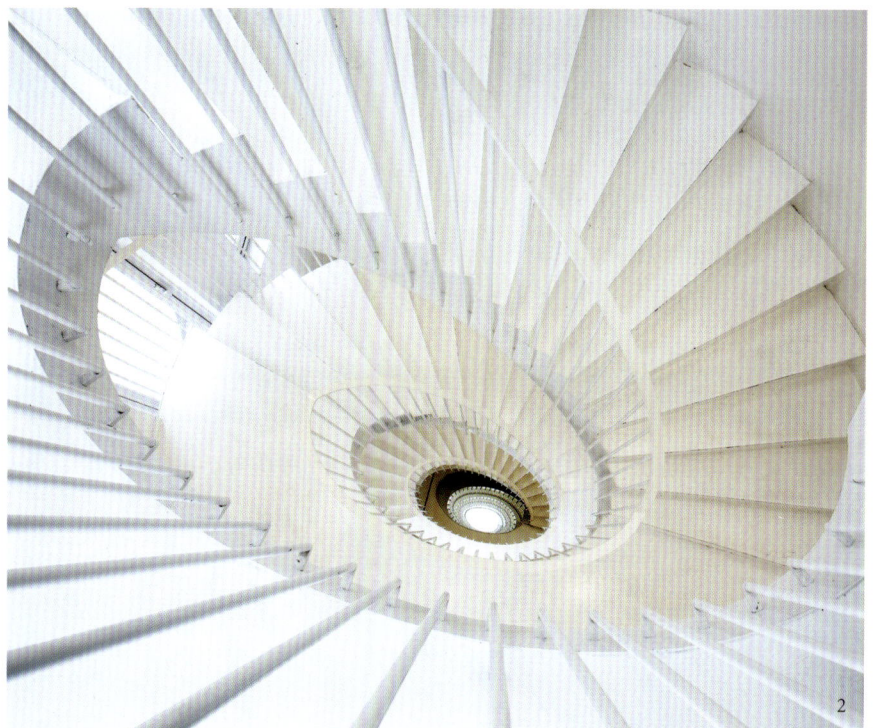

3

4

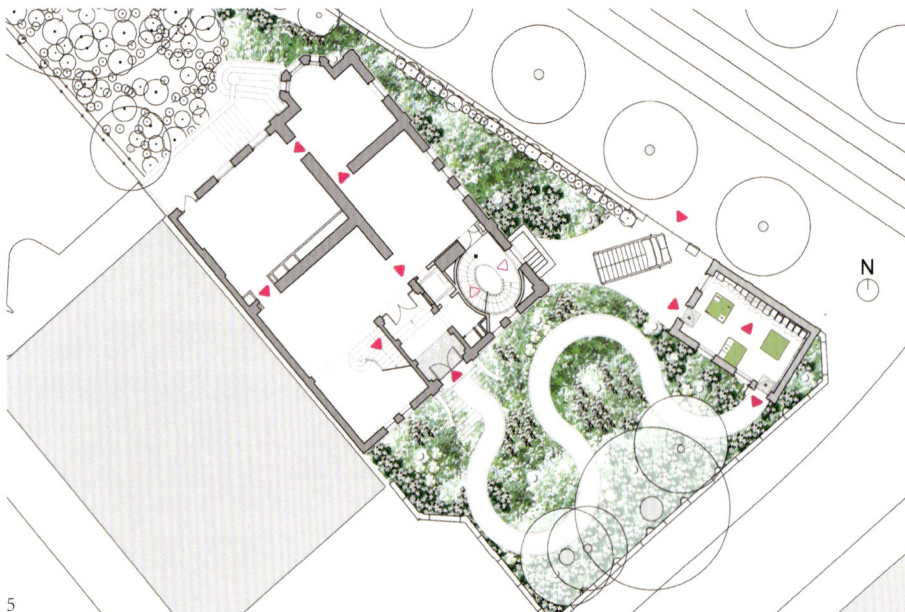

5

a lift or a white spiral staircase forming an internal column to the left of the entrance.

While remaining true to its historical context, the Villa Greiner has acquired a new identity as a cultural centre through weaving its specificity and constraints into the poetry of the whole.

1. Exhibition room, in a "white cube" aesthetic suitable for the display of graphic works
2. Creation of a second interior staircase
3&4. The renovated staircase
5. Plan of the lower ground floor and garden
6. Exhibition room

1. "白色立方体"展览室,适合展示平面设计作品
2. 第二个内部楼梯
3、4. 翻新的楼梯
5. 花园和较低地面层平面图
6. 展览室

汤米·温格尔美术馆和国际插画中心位于新城区的中心,临近协和广场。美术馆坐落在建筑师萨缪尔·雷弗尔于1884年建造的葛来娜庄园内。建筑的新古典主义风格在19世纪末的德国社区十分流行,极具历史价值。建筑师需要在尊重原有设计的前提下,将庄园成功打造成斯特拉斯堡地区的主要文化景点。

这座中产阶级住宅建筑转变了身份,变成了一座公共建筑,欢迎着研究者、学者和普通爱好者的到来。建筑师在注重必要的调节需求和技术事宜的同时,赋予了建筑独特的形象,在完整的建筑环境中融入了文化空间。

项目在19世纪的布景中进行了现代改造。建筑外观的完整性得以保留,唯一的变动在于窗口的改造(窗口必须经过升级来适应美术馆的特殊需求)和花园中蜿蜒小径的设置。

建筑内部的原始房间比例得到了留存,白色的天花板、地面和墙壁让房间像"白色立方体",十分适合展示平面设计作品。参观者可以通过电梯或白色旋转楼梯在美术馆的四层楼之间任意穿梭。位于入口左侧的旋转楼梯形成了一个内部立柱。

通过将特性与约束编织成一首完整的诗歌,葛来娜庄园忠实地保留了自身的历史价值,同时也获得了一个全新的身份——文化中心。

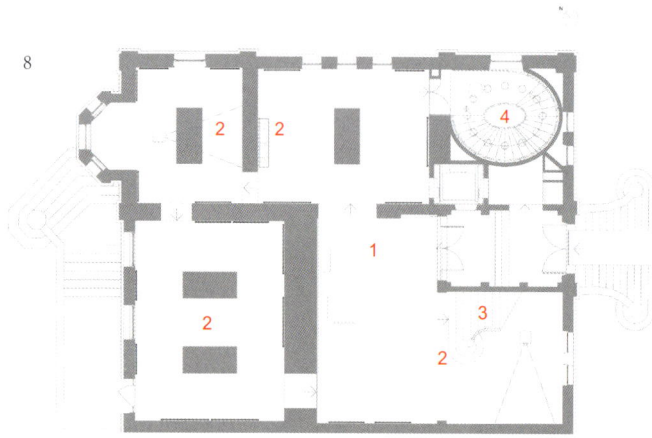

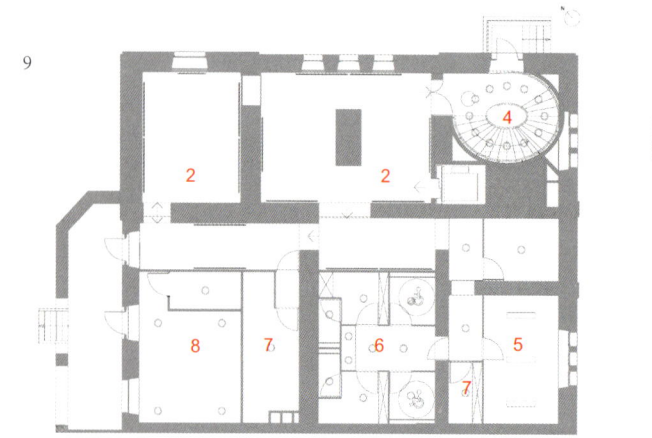

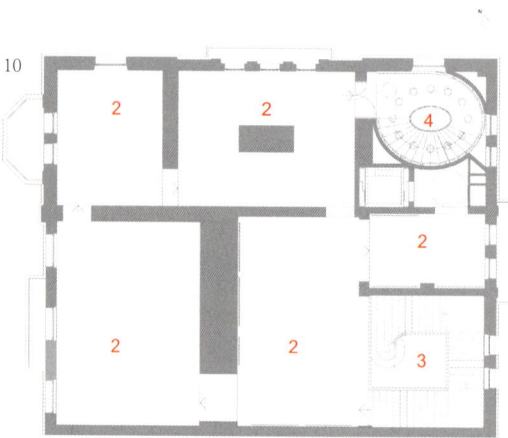

1. Entrance hall
2. Exhibition room
3. Renovated existing staircase
4. New staircase
5. Relaxation room
6. Coat check / WC
7. Maintenance room
8. Technical area

1. 入口大堂
2. 展览室
3. 翻新的楼梯
4. 新楼梯
5. 休息室
6. 寄物处 / 洗手间
7. 维修室
8. 技术区

7&11. Exhibition rooms with white ceilings, floors and walls
8. Higher ground floor plan
9. Lower ground floor plan
10. 1st floor Plan
12. Cross section
13. Landing of the new interior staircase

7、11. 白色天棚、白色地板和白色墙壁的展览室
8. 高地面层平面图
9. 低地面层平面图
10. 二层平面图
12. 横截面
13. 新设置的室内楼梯

11

12

0 1 2 3 4 5

13

DIALOGUES
BETWEEN OLD AND NEW

新旧对话

VALENCE MUSEUM
OF FINE ART AND ARCHEOLOGY

瓦朗斯美术考古博物馆

建筑师 Architect: Jean-Paul Philippon
地点 Location: Valence
完工日期 Completion Date: in progress
图面资料 Visual documents: Michaël Belolo

The museum is housed in the old episcopal palace, facing the cathedral. This site, which has been in constant transformation since the Gallo-Roman period, is like a palimpsest preserving all the imprints of its successive evolutions. Now the 21st century is making its own mark.

The look-out tower over the Rhône is the crowning glory of a "palindrome walk". Situated at the crossroads of two of Valence's historic strata, the project makes the most of the views the building commands across the old city and the 19th-century town.

The focus of the museum is the Hubert Robert collection, where fine art and archeology meet. This painter was a unique figure in the history of art, passionate about ruins and landscapes. By respecting and developing the spirit of the place, the project opens up possibilities for the museum concept to evolve, increasing

1

its appeal for visitors. The impressive collections and the spaces that house them have lent themselves to several different itineraries around the museum: one that is directly chronological, one with a reverse chronology and a third, architectural itinerary. The solidity and complexity of the existing volumes has been preserved, while the new parts adopt the form of successive planes.

2

博物馆设在一座古老的主教宫里，朝向大教堂。项目场地自高卢罗马时代就一直进行着改造，就像是保留了所有进化印记的重写本一样。现在，它正创造着自己的21世纪印记。

面向罗纳河的眺望台是"回文走道"上的高音符。项目坐落在瓦朗斯两个历史地层的交叉口，项目充分注重了建筑所享有的古城和19世纪城镇的全景视野。

博物馆聚焦于休伯特·罗伯特的藏品，汇聚了美术作品和考古文物。这位画家是艺术史上一位独特的人物，他对遗迹和景观有着无比的热情。在尊重并开发场所精神的前提下，项目开创了博物馆概念进化的可能性，增加了它的吸引力。令人瞩目的藏品和收藏空间开辟了多条参观路线：一条以时间为顺序，一条倒叙，另一条则以建筑路线为主。原有空间的实体感和复杂性得到了保留，而新结构则采用了连续的平面造型。

1. Perspective drawing of the museum in its setting
2. The museum, the cathedral and the loop in the Rhône
3. Perspective drawing of the museum entrance on Ormeaux Square
4. Gallo-Roman archeology room

1. 博物馆透视图
2. 博物馆、大教堂和罗纳河
3. Ormeaux广场上的博物馆入口的透视图
4. 高卢罗马考古室

5

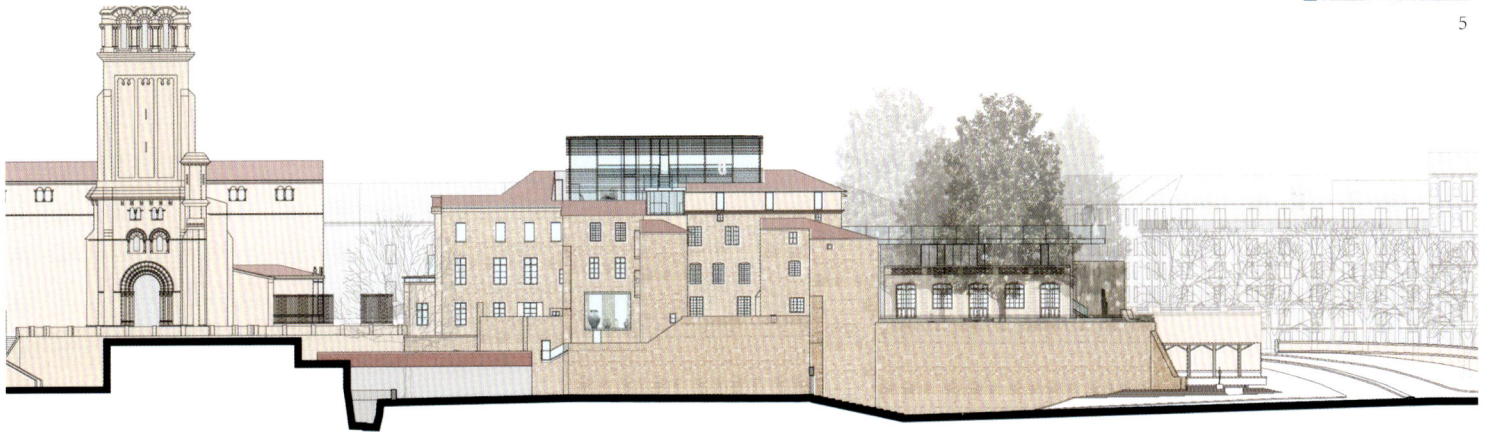

6

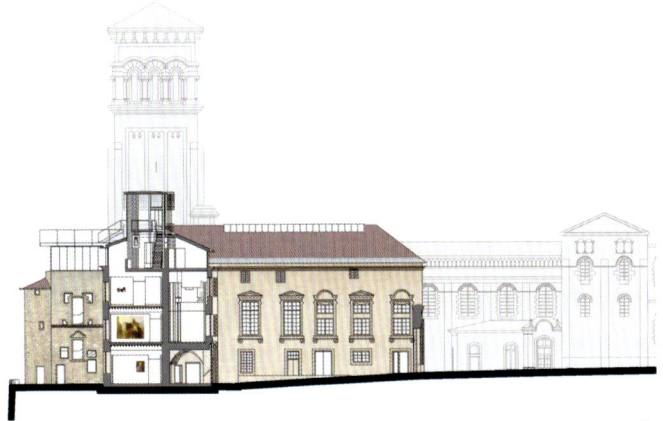

7

5. Façade overlooking the Rhône and look-out tower
6. Museum façade in front of the cathedral
7. Section of the look-out tower and the cour d'honneur
8. Museographical cross section

5. 临罗纳河立面和眺望台
6. 朝向大教堂的博物馆立面
7. 眺望台和荣誉中庭剖面图
8. 博物馆展览场景横向剖面图

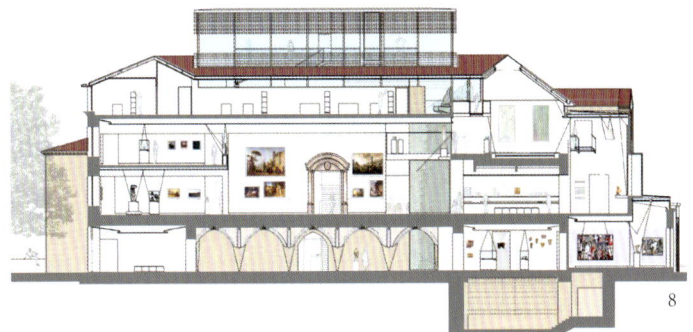

8

QUIMPER MUSEUM OF FINE ARTS

坎佩尔美术馆

建筑师 Architect: Jean-Paul Philippon
地点 Location: Quimper
完工日期 Completion Date: 1993
摄影师 Photographer: Jean-Marie
Monthiers (except n°3)

The current museum, which houses the important Silguy collection and the Max Jacob fund, is an extension and restructuring of the old museum, a neo-Tuscan building designed in 1864 by Joseph Bigot that faces Quimper cathedral. Creating a dialogue between two forms of architecture from different eras, the building has been totally reinterpreted on the theme of split-personality.

The granite staircase has been displaced from the main axis, as if "carried by a wave", leaving the axis, now extended to rue Verdelet, as an opening onto the reception spaces at the heart of the museum. The Fauvist decor of the old Hôtel de l'Épée by Jacques-Julien Lemordant has been reinstated in a transversal wooden nave, floating between two floors. Jean-Paul Philippon's project was based on the principle of transparency and free movement. Brittany's ethereal and changing light enters the building, playing

on the concrete, granite and beech wood.

A majestic central nave, harmonious in its volumes, gives onto the exhibition rooms and all the reception services. Openings in the concrete and granite provide interior and exterior reference points for the visitor, particularly those that reveal a view of the cathedral.

1. Painting rooms
2. Lemordant room
3. Entrance façade on Saint Corentin Square
4-5. Façade on Rue Verdelet
6. Entrance gallery
7. Section

1. 画室
2. 勒莫尔丹特室
3. 圣科朗坦广场的入口立面
4、5. 临维尔德里特街的立面
6. 入口长廊
7. 剖面图

现在，美术馆内收藏着西尔基的重要藏品，也是马克斯·雅各布基金会的办公地点，是旧博物馆（一座由约瑟夫·毕格特于1864年设计的新托斯卡纳风格建筑，正对坎佩尔大教堂）的扩建改造工程。建筑师在两种不同时代的建筑形式之间建立了对话，以分裂的主题重新对建筑进行了诠释。

花岗岩楼梯被从建筑中轴移开，就像"被海浪移开"一样，延伸到维尔德里特街，成为了博物馆中心接待区的开口。重剑酒店由雅克–朱利安·勒莫尔丹特设计的野兽派装饰被复制在横贯的木制中庭，飘浮于两层楼之间。让–保罗·菲里朋的项目以通透感和自由移动原则为基础。布列塔尼缥缈而变换的光线进入楼内，在混凝土、花岗岩和榉木之间跳跃着。

宏伟的中庭拥有和谐的尺度，通往展览厅和所有服务设施。混凝土和花岗岩墙壁上的开口为参观者提供了室内外参照点，特别突出了大教堂的景色。

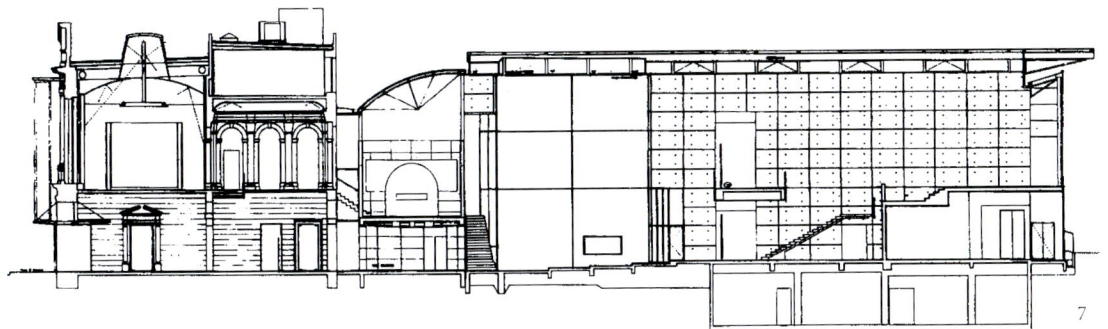

8

9

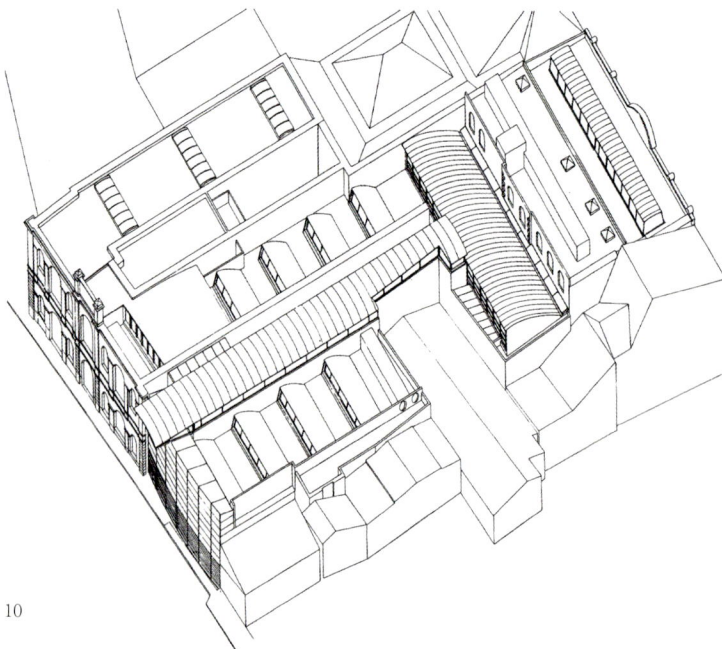

10

8. The nave, lateral view
9. Lemordant room
10. Axonometric drawing
11. Footbridge

8. 侧面视角中的中殿
9. 勒莫尔丹特室
10. 轴测图
11. 人行桥

BONNARD MUSEUM

伯纳德博物馆

建筑师 Architect: Frédéric Ferrero &
Sylvie Rossi
地点 Location: Le Cannet
完工日期 Completion Date: 2011
摄影师 Photographer: Claire Palué

The Villa Saint-Vianney is a fine example of Belle Époque architecture. It was built in 1908 as a private residence, became a guest house and then a hotel until 1990. After being saved from demolition by the state heritage department, it was bought by the town of Le Cannet in 1998 in order to turn it into a museum.

From the beginning, the focus was on preserving the spirit of the place. The project forms part of a larger development of the old town centre of Le Cannet, and is linked to Bonnard's former home, the Villa Le Bosquet, by a walk in the steps of the painter. The renovation of the Villa Saint-Vianney develops almost 890m^2 of useable space, including 495m^2 in the existing building. An extension has been built on the sloping ground in front of the villa, containing the reception area, shop, educational room and a large terrace. The façade opens onto the Boulevard Sadi Carnot and marks the entrance to the museum.

A glass column containing a staircase and a lift, connecting to the existing building via a footbridge, takes visitors to the exhibition floors while offering a view over the landscape. The internal structure of the Villa Saint-Vianney has been entirely altered. Air conditioned throughout, the museum unfolds over five floors including three floors of exhibitions, a projection room and a terrace overlooking the garden.

圣维雅纳庄园是美好年代（1900-1914）建筑的典范。庄园建于1908年，最初作为私人住所，随后直到1990年，一直都是宾馆和酒店。在经过国家遗产部门的挽救，免受拆除之后，勒内卡镇在1998年买下了它，试图将它改造成一座博物馆。

从一开始，改造的焦点就在于留存地方的精神。项目是勒内卡古镇中心开发的一部分，通过台阶与伯纳德的故居——丛林庄园相连。圣维雅纳庄园的翻修面积达890平方米，包括495平方米的原有建筑改造。庄园前方坡地上的新建部分内设有接待区、商店、教学室和大型露台。建筑朝向萨迪卡诺大道，标志着博物馆的入口。

巨大的玻璃柱内设有楼梯和电梯，通过天桥与原有建筑相连，将参观者带到展览区，同时也展示了外面的风景。圣维雅纳庄园的内部结构进行了彻底的调整。空调开放的博物馆共分为五层，其中包括三层展览区、一个放映室和一个俯瞰花园的露台。

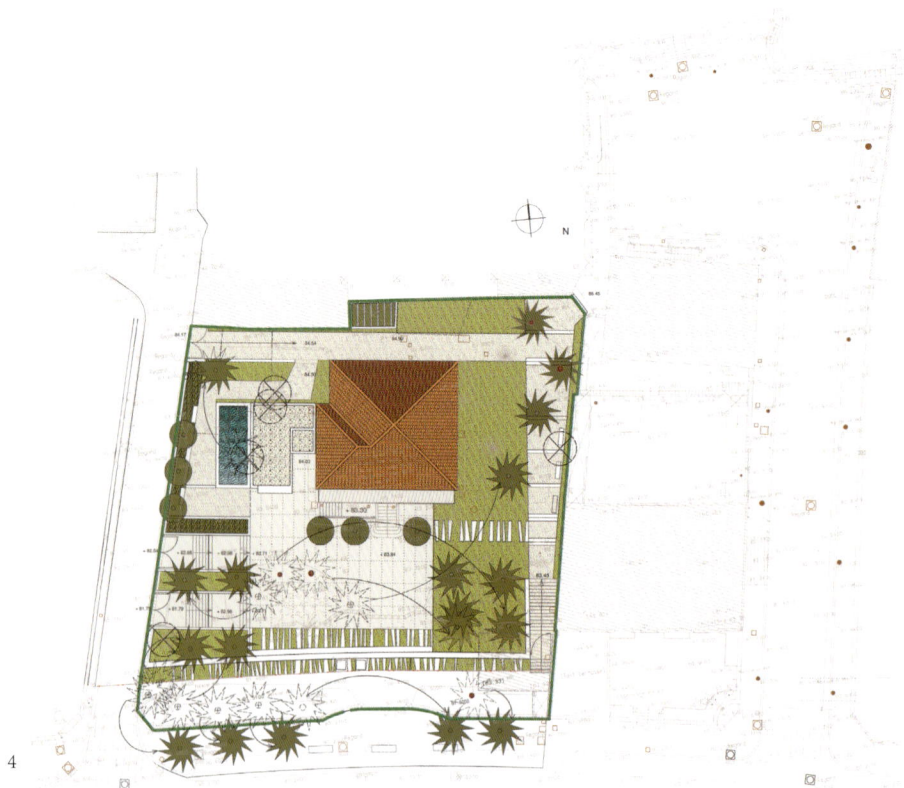

1. Exhibition room
2. Museum entrance, Boulevard Sadi
Carnot
3. View of the museum from Boulevard
Sadi Carnot
4. Master plan
5. View from the terrace
6. South façade
7. West façade

1. 展览厅
2. 面朝萨迪卡诺大道的博物馆入口
3. 萨迪卡诺大道一侧外景
4. 总体规划图
5. 露台外景
6. 南立面
7. 西立面

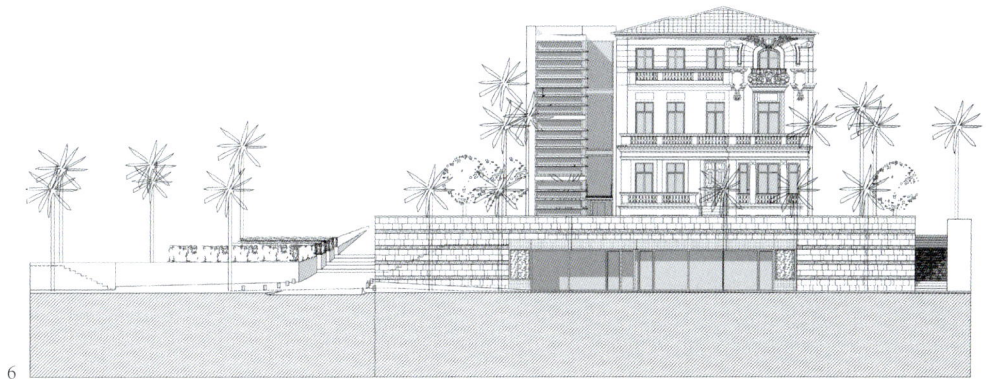

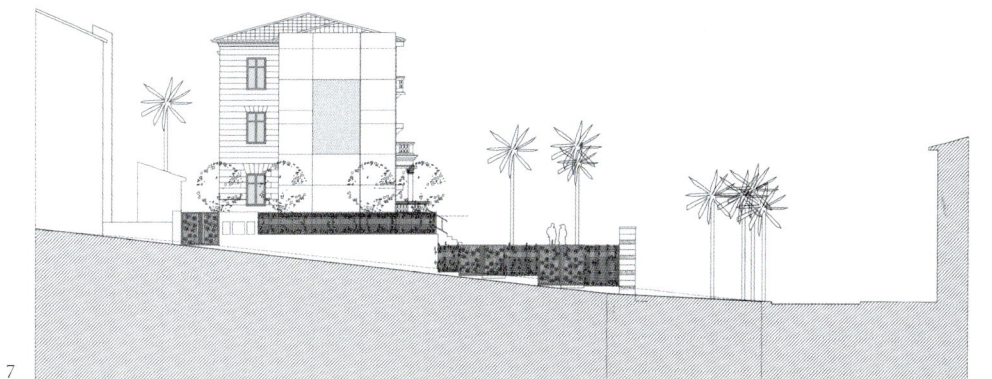

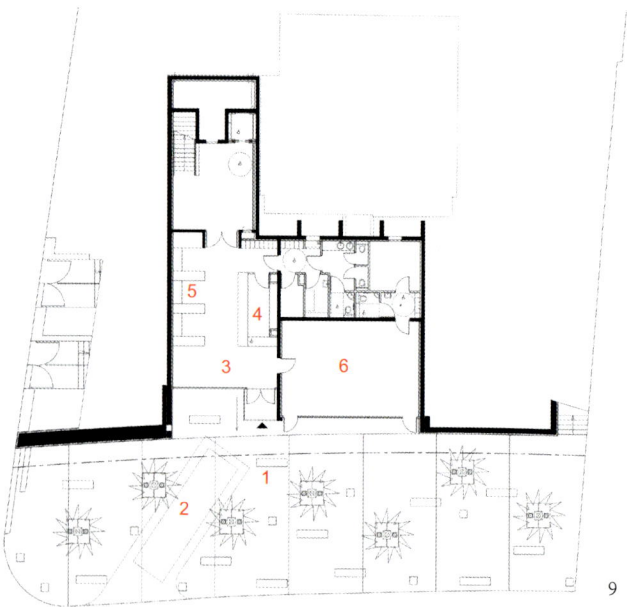

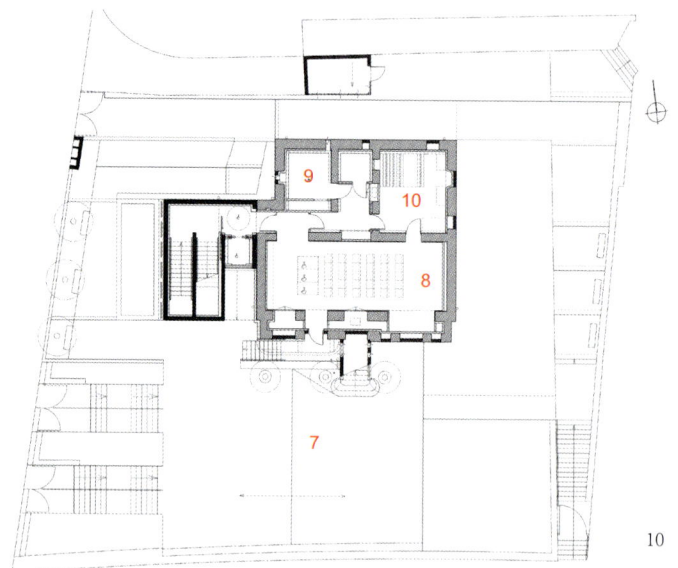

8. The museum's garden terrace
9. Ground floor plan
10. 1st floor plan

8. 博物馆的花园露台
9. 一层平面图
10. 二层平面图

11

12

11. Exhibition room
12. Old staircase of the Hôtel Saint-Vianney
13. 2nd floor plan
14. 3rd floor plan
15. 4th floor plan
16. Roof plan

11. 展览室
12. 圣维雅纳饭店的旧楼梯
13. 三层平面图
14. 四层平面图
15. 五层平面图
16. 屋顶平面图

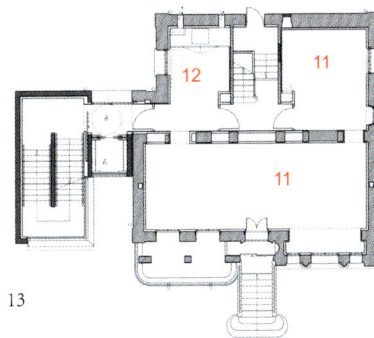

13

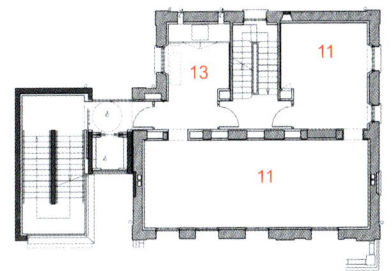

14

1. Forecourt
2. Delivery area
3. Entrance Hall
4. Reception
5. Shop
6. Activities room
7. Garden terrace
8. Projection room
9. Technical area
10. Stock room – reserves
11. Exhibition room
12. Introduction room
13. Biography room
14. Workshop
15. Roof terrace

1. 前院
2. 送货区
3. 入口大堂
4. 前台
5. 商店
6. 活动室
7. 花园露台
8. 放映室
9. 技术区
10. 库存室
11. 展览室
12. 介绍室
13. 传记室
14. 工作室
15. 屋顶露台

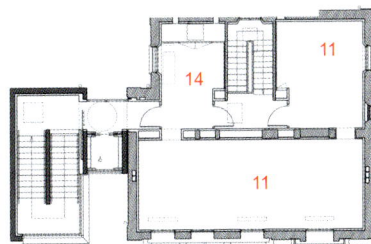

15

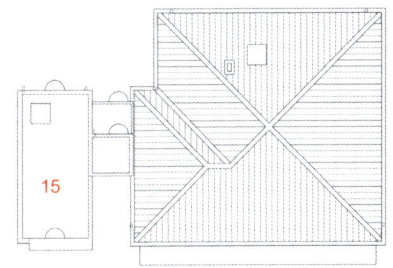

16

COURBET MUSEUM
库尔贝博物馆

建筑师 Architect: Ateliers 234
地点 Location: Ornans
完工日期 Completion Date: 2011
摄影师 Photographer: Nicolas Waltefaugle

The Courbet Museum has left the limited space of its original location in the Maison Hébert to expand into two adjoining buildings: the Maison Borel and the Hôtel Champereux. These three entities rejoice in a remarkable location right beside the River Loue in Ornans, between the Great Bridge, the fountain and Robert Fernier Square. The building has been opened up to the light, colours and materials of "Courbet country".

The previous façade gave no indication of the museum's presence. Today a large stainless steel box juts out, suspended from the height of the entrance hall, announcing the museum to the city. On the river side, a horizontal sequence also proclaims the building's new dimensions: a thin glass roof built into the tiles highlights the position of the hall, then a glassed-in gallery forms a link between the houses, continues through the Maison Hébert and protrudes from the end wall. This long horizontal sequence on the river symbolises the new ambition of the

museum. At nightfall, this line becomes a luminous thread between the houses and the garden reflecting on the water.

One of the basic tenets of the project was to preserve the atmosphere of the Maison Hébert. The parts of the museum that called for large-scale restructuring, such as the welcome area, are kept "at a distance" and connect with the public space. This house, a succession of historic rooms of which some are heritage listed, is plunged into the atmosphere and colours

1. The museum's new dimensions; a luminous thread
2. The old courtyard houses the entrance hall. The black box metaphorically reaches out to the town
3. The black box floats above the reception
4. The gallery, a place of immersion into the landscape
5. Footbridge to the black box
6. Situation plan

1. 博物馆新造型，一条光带
2. 入口大堂。夜光下像黑盒子延伸到小镇
3. 飘浮在前台上空的黑盒子
4. 长廊
5. 人行桥
6. 方位图

of Courbet's time. The Maison Hébert leads into the Revolutions rooms, where the scene is first set in space and time: a display case/gallery visible from both sides evokes the revolution of 1848 on one side and on the other, at the end of the visit, the revolution of 1870. Thanks to this transparency the building is readable through its entire depth.

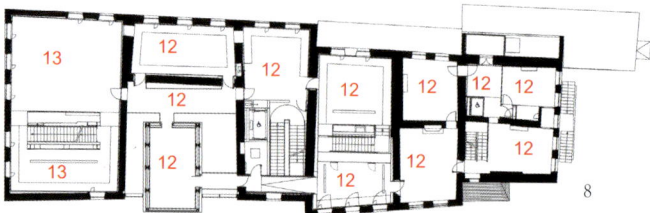

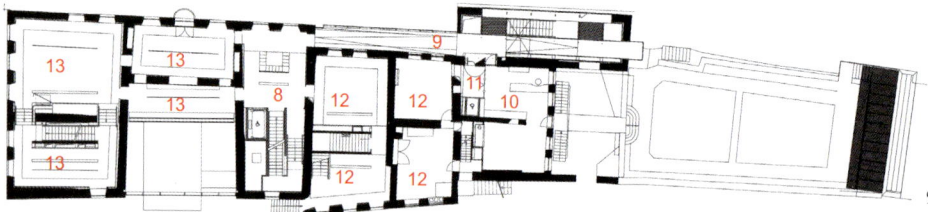

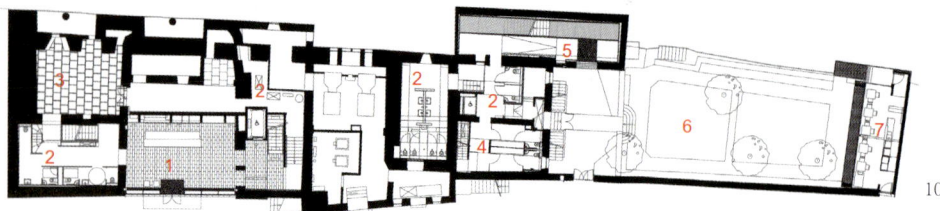

1. Entrance hall / reception
2. Rest rooms
3. Vaulted cellars
4. Staff cloakrooms
5. Gallery, exit to the garden
6. Garden
7. Summer cafeteria
8. Coat check
9. Gallery overlooking the Loue
10. Shop
11. Entrance to the permanent exhibition
12. Permanent exhibition
13. Temporary exhibition

1. 出口大堂 / 前台
2. 休息室
3. 拱形酒窖
4. 员工衣帽间
5. 通往花园的长廊
6. 花园
7. 夏季咖啡馆
8. 存衣处
9. 远眺卢厄河的长廊
10. 商店
11. 常设展区入口
12. 常设展区
13. 临时展区

7. Revolutions room, a place of rupture
8. 1st floor plan
9. Upper ground floor plan
10. Lower ground floor plan
11. The walls cut at an angle show their historic thickness
12. Display tables protect precious drawings

7. 革命厅，空间和时间转变
8. 二层平面图
9. 高地面层平面图
10. 低地面层平面图
11. 墙壁切开一角以显示建筑的历史深度
12. 保护珍贵图画的展示桌

库尔贝博物馆从埃贝尔庄园有限的空间扩展至两座相邻的建筑——波莱尔庄园和香佩罗酒店。这三座建筑沿着奥尔南镇的卢厄河，坐落在大桥、喷泉和罗伯特·菲尼尔广场之间。建筑的光线、色彩和材料都展现了"库尔贝乡村"特色。

之前的外立面丝毫没有显露博物馆的功能。现在一个巨大的不锈钢盒子向外突出，从门厅上方悬吊下来，向城市宣告着博物馆的存在。河畔的水平序列同样宣示着建筑的新造型：嵌入瓷砖的薄玻璃屋顶突出了大厅的位置，而玻璃走廊则连接了住宅，穿过埃贝尔庄园，从端墙伸出。河畔长长的水平序列象征着博物馆的雄心。黄昏，这条线在住宅和花园之间形成了一条闪烁的光带，倒映在水面。

从根本上来说，项目旨在保护埃贝尔庄园的氛围。博物馆的一部分需要大规模的重建——例如迎宾区。这些重建部分庄园保持了一定的距离，并且与公共区域相连。这座列入遗产保护名录的住宅已经融入了库尔贝时期的氛围和色彩。埃贝尔庄园通往革命厅，展厅的布景首先以空间和时间开端：展示走廊的两侧分别让人想起了1848年的革命和1870年的革命。这种通透感让人们可以轻易了解建筑的整个深度。

15

16

13. The passage from one house to another, in serenity
14. Restrooms in the old vaulted cellars
15. The exit hall, a place of reflection(s)
16. Near the museum exit visitors can look down on the water
17. Cross section
18. Longitudinal section

13. 从一屋到另一屋的通道
14. 位于昔日拱形酒窖的卫生间
15. 光影叠射的出口处
16. 近出口处游客能够低头看到河水
17. 横截面
18. 纵剖面

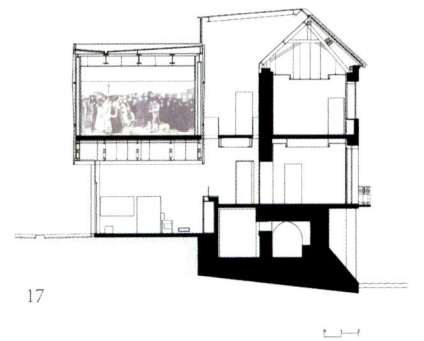
17

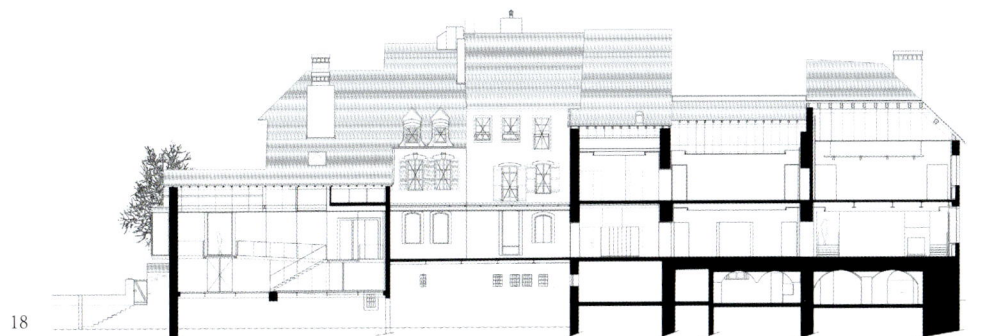
18

LA PISCINE
MUSEUM OF ART AND INDUSTRY
游泳池艺术与工业博物馆

建筑师 Architect: Jean-Paul Philippon
地点 Location: Roubaix
完工日期 Completion Date: 2001
摄影师 Photographer: Florian Kleinefenn

The Museum of Art and Industry has given a new dynamism to the town of Roubaix, which was proud of its old Art Deco swimming pool designed by Albert Baert. This building has been redeveloped and extended into a neighbouring disused industrial site, thereby offering the potential for cross-fertilisation with the worlds of industrial creation (fashion shows, textile design), culture (performance arts, music) and education (teaching workshops).

The project has made use of the original materials, bricks and enamelled stoneware tiles of the cabins, found in situ and has reconfigured them. Through a dialogue that respects the existing building, down to the very detail of the its vocabulary, this project has given a central place to creative mutations and scenographies that are the height of creativity. The thin stream of water preserved in the central swimming pool area reflects the works presented on the horizontal pontoons and can also be covered

over for an exhibition or a fashion show.

With an auditorium, and logistical spaces that are entered from the street, the vast temporary exhibition space enables extremely varied exhibitions (Picasso, Degas, Signac, Marimekko) that make use of a large-scale hanging system that is both fixed and mobile. It is a new kind of museum that brings together art and economics, memory and modernity.

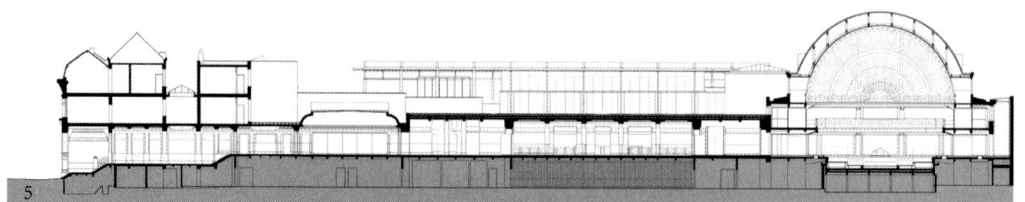

1. Exterior view
2. Façade of the museum
3. The entrance through the garden on the old industrial site
4. Master plan
5. General section
6. Entrance hall

1. 博物馆外观
2. 博物馆沿街立面
3. 由旧工业场地整治成的花园成为博物馆的迎宾入口
4. 总体规划
5. 总剖面
6. 入口大堂

7&8. Entrance to the temporary
exhibition
9. Section of the entrance hall

7、8. 通往临时展区的入口
9. 入口大堂剖面图

艺术与工业博物馆赋予了鲁贝镇全新的活力。此前，鲁贝镇一直以艾伯特·贝尔特所设计的装饰艺术游泳池而自豪。建筑经过重新开发，扩展至附近一个废弃的工业场地，提供了工业（时装秀、纺织品设计）、文化（表演艺术、音乐）和教育（教学工坊）跨界互动的机会。

项目利用了原始材料——砖块和小屋里的瓷砖，对它们进行了重新配置。在尊重原有建筑每个细节的基础上，项目将重点放在了创造性改造和透视布景中。中央泳池

区的潺潺细流倒映出浮桥上的展示的作品，也可以被覆盖起来进行展览或时装秀。

宽敞的临时展览空间内配有礼堂和物流空间（可直接从街道进入），适合各种各样的展览配置（毕加索、德加、西涅克、玛丽梅科）。大幅的悬挂系统可固定也可移动。这是一种新型博物馆，它将艺术与经济、回忆与现代融合在一起。

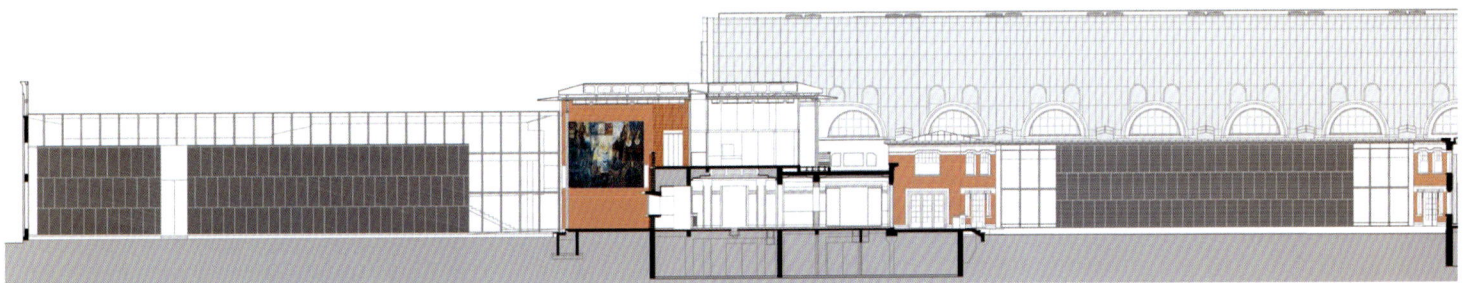

10. The swimming pool
11. Cross section of the swimming pool
12. Longitudinal section of the swimming pool
13. The swimming pool, sculpture pontoon and changing rooms

10. 游泳池
11. 游泳池横截面图
12. 游泳池纵切面图
13. 游泳池、雕塑浮桥和更衣室

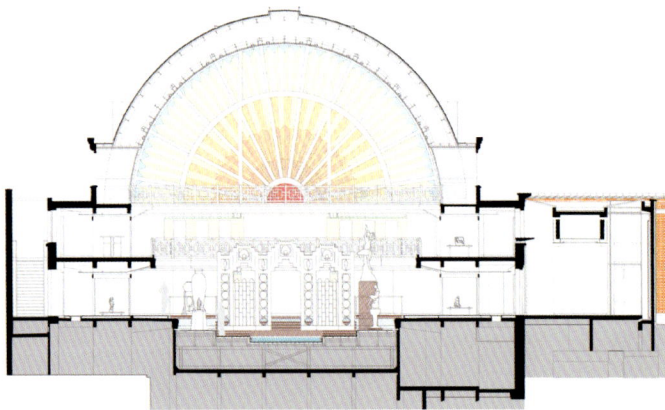

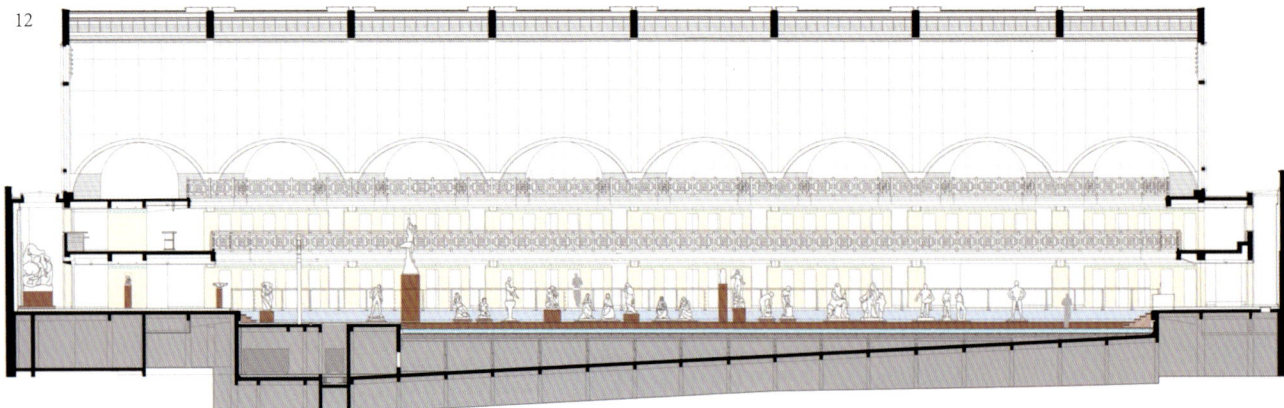

14

14. A Carolyn Carlson performance in the
swimming pool
15. Entrance through the compass room
and the bust of Albert Baert

14. 游泳池中的卡罗琳·卡尔森表演
15. 通向指南针室和艾伯特·贝尔特半身像的入口

LA MAISON ROUGE
红色之家

建筑师 Architect: Amplitude
Architectes, Jean-Yves Clément
地点 Location: Paris
完工日期 Completion Date: 2004
摄影师 Photographer: Luc Boegly

Originally, the Maison Rouge (the Red House) was a shop open to the street. The back wall of the shop was demolished, opening up a courtyard, in the centre of which was found a small house encircled by a glass roof and surrounded by warehouses from the beginning of the 19th century, whose riveted metal beams retained a historic value. Behind were two industrial buildings, one from the 1950s-60s and the other in concrete, over three floors. Jean-Yves Clément's idea was to remove the floors of the latter to create a large volume.

The heterogeneity of this site has become one of its strengths. It has been used according to the specificity of each place within. Some spaces are large on the lower part, like the current polygonal room. Further on, the architects find what is known as a "White Cube", and down a few steps you feel as if you are under a tent where the nave of the original industrial structure has been preserved. The idea was that on entering you could

forget that you were in Paris and lose any geographical notion of the city. For Clément it was about offering a place that could evoke certain buildings from other large capitals, which would aid the appreciation of contemporary art.

The false ceilings that in the past unified the spaces and made this into a typically Parisian building have been avoided. For the restaurant, the idea was that it would be discovered, in a democratic way, at the heart of the Maison Rouge. Anyone can

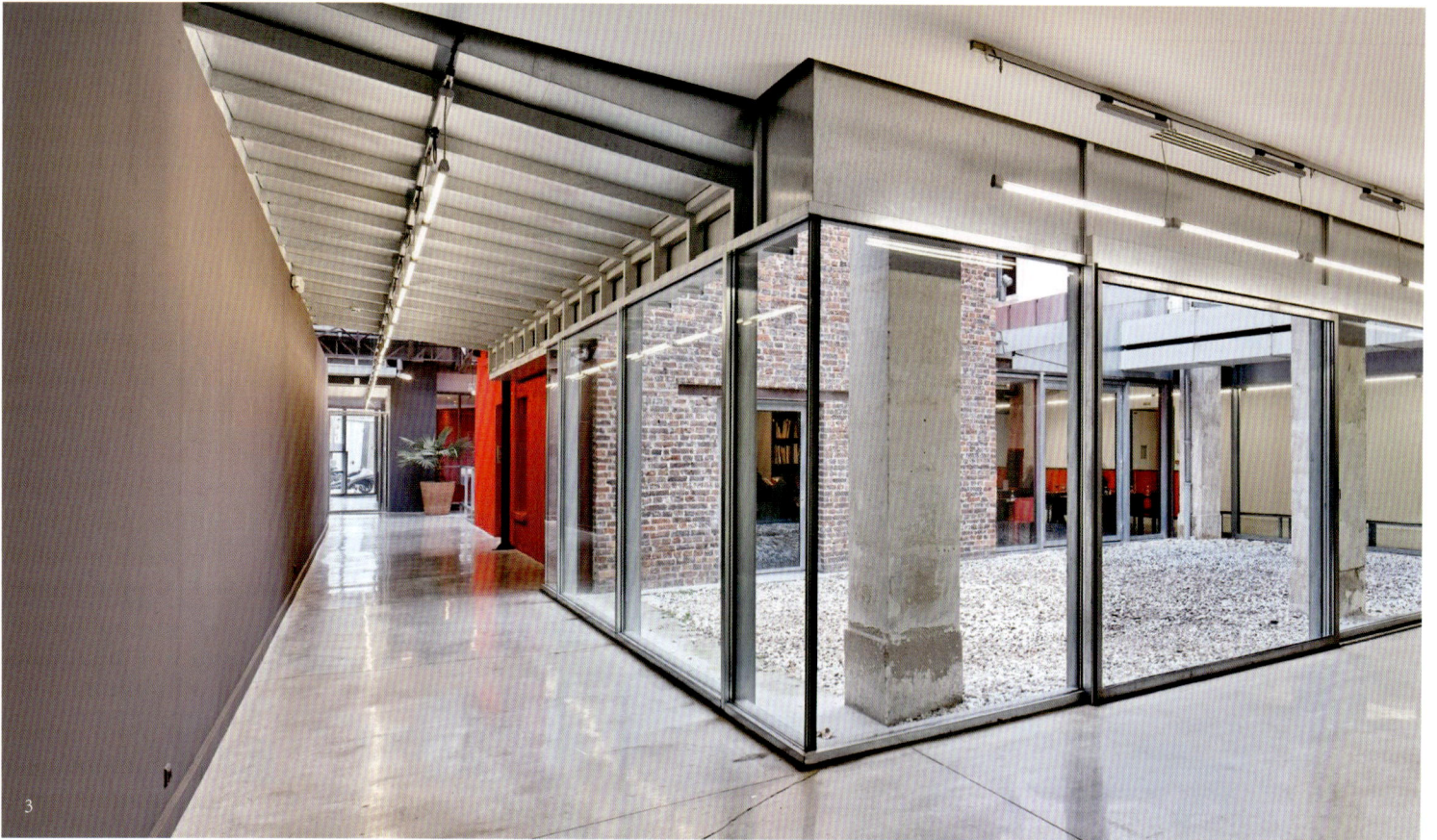

1. Gallery, patio and café
2. Skylight
3. Gallery with a view of the patio
4. Upper mezzanine gallery
5. Mezzanine

1. 长廊、庭院和咖啡厅
2. 天窗
3. 带有庭院一角的长廊
4. 夹层上的长廊
5. 夹层

enter, without a ticket, and have a cup of coffee at the heart of the exhibition. The city and the public space are thus invited in to the private foundation. This kind of neutrality of spaces is what distinguishes the Maison Rouge from other Parisian environments. We are nowhere and we could be anywhere… Here, in this non-referential space and time, contemporary art meets post-industrial architecture.

最初，红色之家是一家临街的商店。店铺的后墙被拆除，通往庭院，而庭院的中央是一件由玻璃屋顶包围的小屋，四周环绕着仓库。仓库建于19世纪初期，其铆接金属梁具有历史价值。庭院后方是两座工业建筑，一座建于20世纪50、60年代，另一座采用混凝土结构，共有三层。让·伊夫·克莱芒的想法是移除后者的楼层，以建造更大的空间。

场地的多样性反而成为了它的长处，每个场所都物尽其用。一些空间的底部宽敞，就像多边形房间一样。一些空间像"白色立方体"，而工业建筑的中庭则让人觉得像置身于帐篷之中。走进博物馆，你会忘记自己是在巴黎，丧失一切关于这座城市的地理概念。克莱门特希望博物馆能够让人想起其他的大都市，帮助参观者埋解现代艺术的精髓。

建筑师避免了在博物馆设计中使用巴黎建筑将各个空间统一起来的假天花板。餐厅设在红色之家的中心，十分显眼。无需买票，人人都能进入餐厅，边享用咖啡边参观展览。这样一来，城市和公共空间被引入了私人空间之中。这种空间中立性让红色之家与巴黎的其他环境区分开来，既是无名之所，又是天涯海角。在这个没有指向性的时空中，现代艺术与后工业建筑欣然邂逅。

6

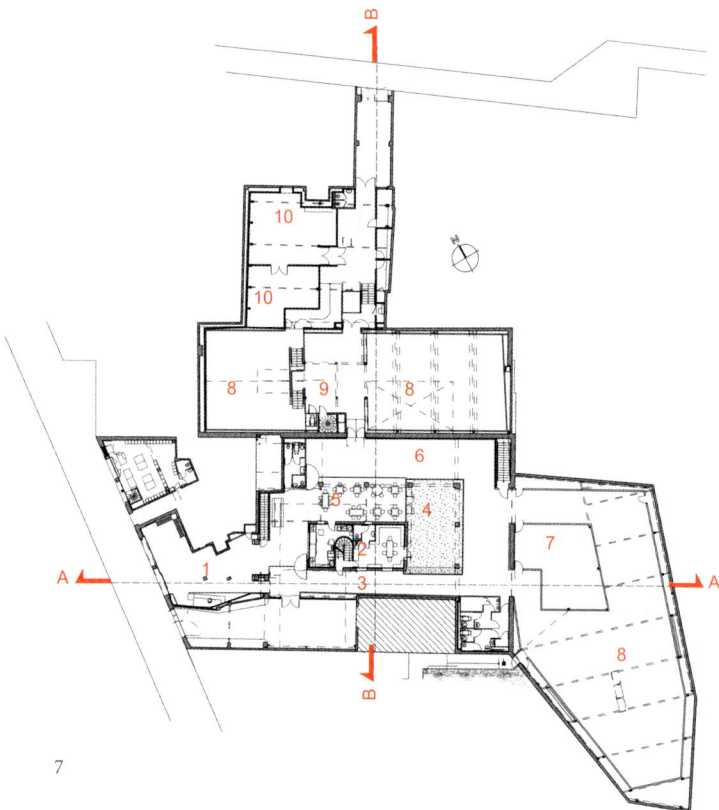

7

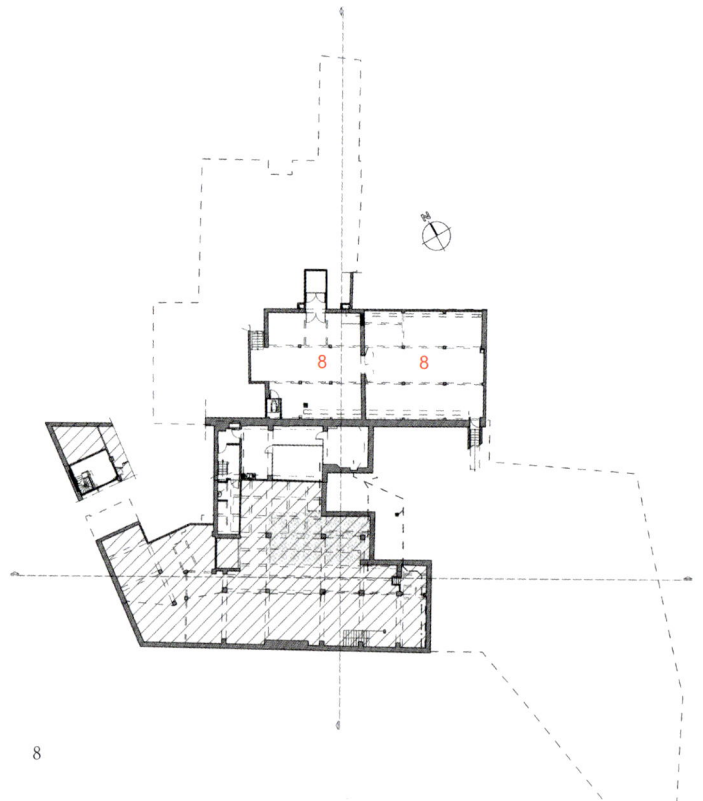

8

1. Entrance hall
2. Maison Rouge
3. Interior street
4. Patio
5. Terrace café
6. North gallery
7. Multimedia room
8. Exhibition room
9. Balcony
10. Reserves

1. 入口
2. 红色之家
3. 室内街道
4. 中庭
5. 咖啡厅
6. 北边展廊
7. 多媒体展示室
8. 展览空间
9. 休息平台
10. 储藏空间

6. Polygonal room
7. Ground floor plan
8. Basement floor plan
9. Section AA
10. Section BB

6. 多边形房间
7. 一层平面图
8. 地下室平面图
9. AA剖面图
10. BB剖面图

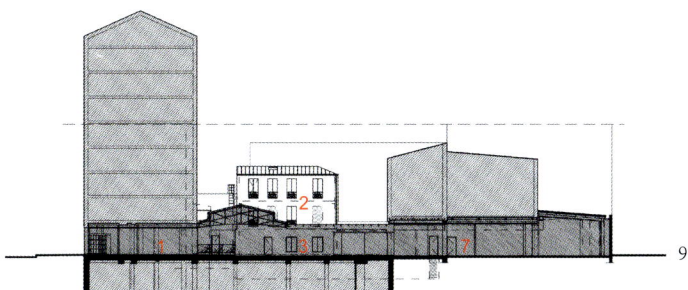

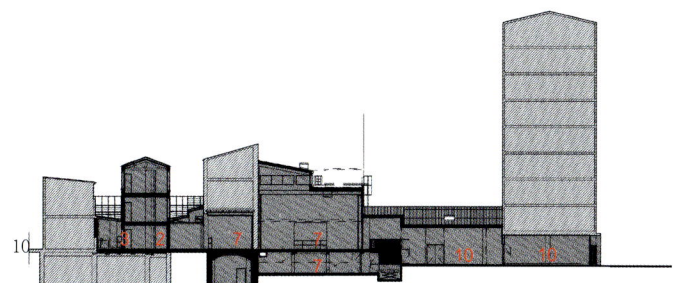

MATISSE MUSEUM

马蒂斯博物馆

建筑师 Architect: Beaudouin-Husson
地点 Location: Cateau-Cambrésis
完工日期 Completion Date: 2002
摄影师 Photographer: Jean-Marie
Monthiers

Architecture and painting follow paths that converge at certain points. For both, there is no difference between the inside and the outside. The space is unitary, it always stays the same, no screen separates it into two, it remains as one.

The Matisse Museum expresses this continuous character of space. It is also an attempt to unite in a single form fragments of buildings and landscapes that, through their history and their situation, are close in space but far away in time: a square,

a palace, a school, a park. The aim was to turn these elements into a united whole, a space where each part would be indispensable to the balance of the whole.

The architects sought harmony in the proportions of the museum, giving a musical resonance to the whole. The architecture is based on the superimposition of elements, on the progressive disappearance of what lies behind the things we see. The architecture of a museum is like a machine to slow

down time, it forces one to adjust from the accelerated time of everyday life to the more tranquil time of contemplation. The association of two invisible laws of nature makes light and gravity more present. In architecture, light can have weight. If you give weight to light its movement slows down, the light becomes solemn, it settles into a slow thickness. The architecture of the museum is designed as a resonant casing in which this emotion vibrates.

建筑和绘画的路线在某一点会合，二者都没有内外的区别。空间是一个不变的统一体，没有什么能将它分割，它始终是一个整体。

马蒂斯博物馆展现了空间的连续性。它尝试以单一的形式把建筑和景观的片段统一起来。基于历史和地理的缘由，这些片段在空间上十分靠近，在时间上却十分遥远：广场、宫殿、学校、公园。设计的目标是将这些元素转化为统一的整体，在这个整体中，每个部分都是不可或缺的。

建筑师在博物馆的比例中寻求和谐，塑造一个协调共鸣的整体。建筑以各种元素的重叠为基础，让隐藏在我们所看见事物背后的意义逐渐消失。博物馆建筑就像一台让时间变缓的机器，它带领人们从日常中被加速的时间进入到凝赏与冥想中的沉静时光。光和重力这两个大自然无形法则的结合使它们自身变得更加明显。在建筑中，光也有重量。如果为光加重，它就走得更慢，变得肃穆而厚重。博物馆建筑的设计正是为了引起这种情绪震动的共鸣。

6

1. Drawings and sculptures pavilion
2. New wing
3. Cafeteria
4. Matisse room
5. Master plan
6. Sculpture room
7. Section and west façade

1. 绘画和雕塑馆
2. 新翼楼
3. 咖啡馆
4. 马蒂斯厅
5. 总体规划图
6. 雕塑厅
7. 剖面和西立面图

7

8 9

10 11

12 13

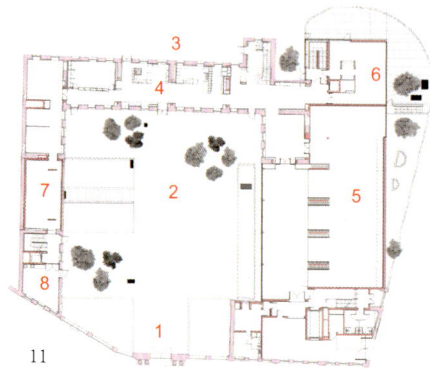

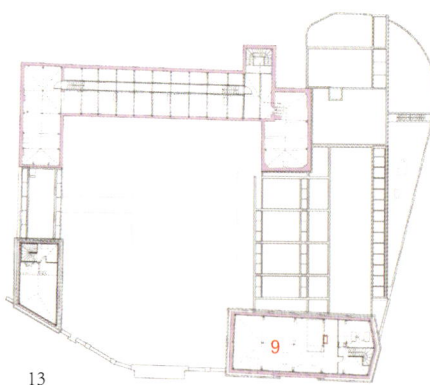

8-9&14. Matisse gallery on the 1st floor
10. Basement floor plan
11. Ground floor plan
12. 1st floor plan
13. 2nd floor plan

8、9、14. 马蒂斯画廊二层
10. 地下室平面图
11. 一层平面图
12. 二层平面图
13. 三层平面图

1. Entrance	1. 博物馆入口
2. Courtyard	2. 中庭
3. Park	3. 公园
4. Reception / bookshop	4. 接待厅 / 书店
5. Temporary exhibitions	5. 临时性展览空间
6. Cafeteria	6. 咖啡厅
7. Drawings room	7. 图画馆
8. Educational services	8. 教学空间
9. Administration	9. 行政空间
10. Herbin rooms	10. Herbin 捐赠作品
11. Tériade Collection	11. Tériade 出版艺术图书
12. Concrete Art	12. 具象艺术
13. Matisse rooms	13. 马蒂斯室
14. Sculptures	14. 雕塑作品
15. Auditorium	15. 演映厅
16. Reserves	16. 储藏空间

14

LANGUEDOC-ROUSSILLON
REGIONAL MUSEUM OF CONTEMPORARY ART

朗格多克–鲁西永地区现代艺术博物馆

建筑师 Architect: Pierre-François Moget & Anne Gaubert, Projectiles
地点 Location: Sérignan
完工日期 Completion Date: 2006
摄影师 Photographer: Jean-Paul Planchon

In the heart of the Languedoc-Roussillon region, beside the Mediterranean in the small town of Sérignan, the Regional Museum of Contemporary Art (MRAC) is situated in an old wine-makers' warehouse.

The project for transforming it into a museum was entrusted to the architects Pierre-François Moget and Anne Gaubert in 2005-2006. This development provided the region with 2,700m^2 to show temporary exhibitions of its collections and to offer an itinerary through different spaces including the graphic arts room, the video space, the children's workshops and a book and gift shop.

Since 2010, it has been enlarged with a new 500m^2 platform for experimental projects. On the ground floor, large

volumes, alternating the more intimate, ceilinged rooms with those whose beams are exposed, are all bathed in natural light. The Projectiles workshop created all the museum furniture including the cabinets in the graphic arts room. A real museum within the museum, the graphic arts room, plunged into half-light, offers the visitor a new experience of perception.

With his work *Rotation,* the artist Daniel Buren has encircled the museum with a belt of colour applied to all the windows, creating visual effects both on the interior and the exterior. A large ceramic fresco, *Les Femmes Fatales*, by the artist Erró, adorns the exterior façade.

1. Exterior view
2. Erró, *Les Femmes Fatales*, 1995-2006
3. Projections of *Rotation*, installation by Daniel Buren
4. Entrance hall, with a work by Lawrence Weiner on the atrium walls
5. Master plan
6. "Sémiose Éditions" exhibition
7. The museum collection

1. 博物馆外观
2. 由艺术家埃罗创作的大型陶瓷壁画"致命女子"装饰着建筑的外立面,1995–2006
3. 艺术家丹尼尔·布朗的作品"循环"
4. 入口大厅,中庭的墙壁上有劳伦斯·韦纳的作品
5. 总规划图
6. "Semiose出版品"展览
7. 博物馆内的收藏品

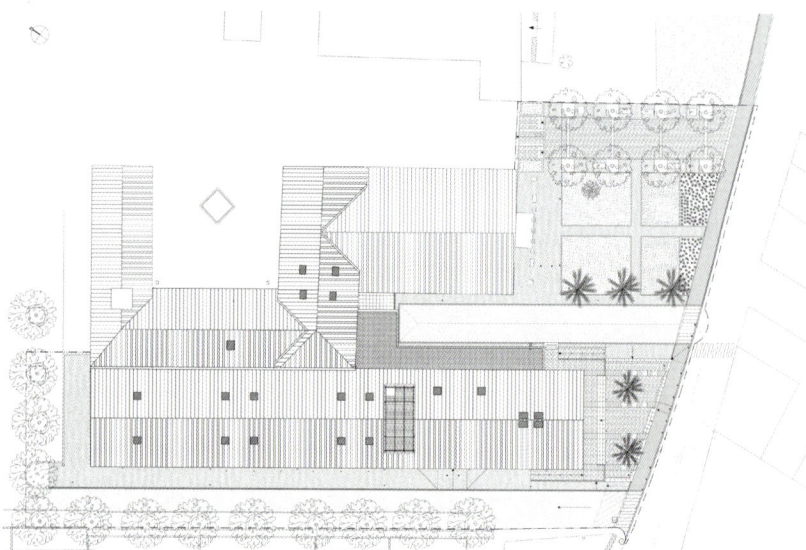

5

8-10. Graphic art

8-10. 平面艺术区

朗格多克－鲁西永地区现代艺术博物馆位于该地区中心的塞里尼小镇，紧邻地中海。博物馆就设在镇上一座古旧的酿酒仓库中。

博物馆改造项目由建筑师皮埃尔·莫戈和安妮·高博特在2005－2006年期间完成。项目为当地提供了2,700平方米的空间来进行临时展览，包括平面艺术厅、视频区、儿童工坊和书籍礼品店。

2010年，博物馆为实验项目新增了500平方米的平台。一楼的大空间、裸露横梁的吊顶房间都沐浴在自然光线之中。投射工作室为博物馆打造了全套家具，包括平面艺术厅中的陈列柜。暗光中的平面艺术厅堪称博物馆中的博物馆，为参观者带来了独特的感官体验。

艺术家丹尼尔·布朗用自己的作品"循环"（Rotation）包围了博物馆：彩带被运用到所有窗口上，在室内外都形成了独特的视觉效果。由艺术家埃罗创作的大型陶瓷壁画"致命女子"（Les Femmes Fatales）则装饰着建筑的外立面。

11. "Géographies du Dessin" exhibition
12&13. Hans Hartung, "Spray" exhibition, 2010
14. Sections

11."绘画地理学"展览
12、13. 汉斯·哈通的"喷雾"展览，2010年
14. 剖面图

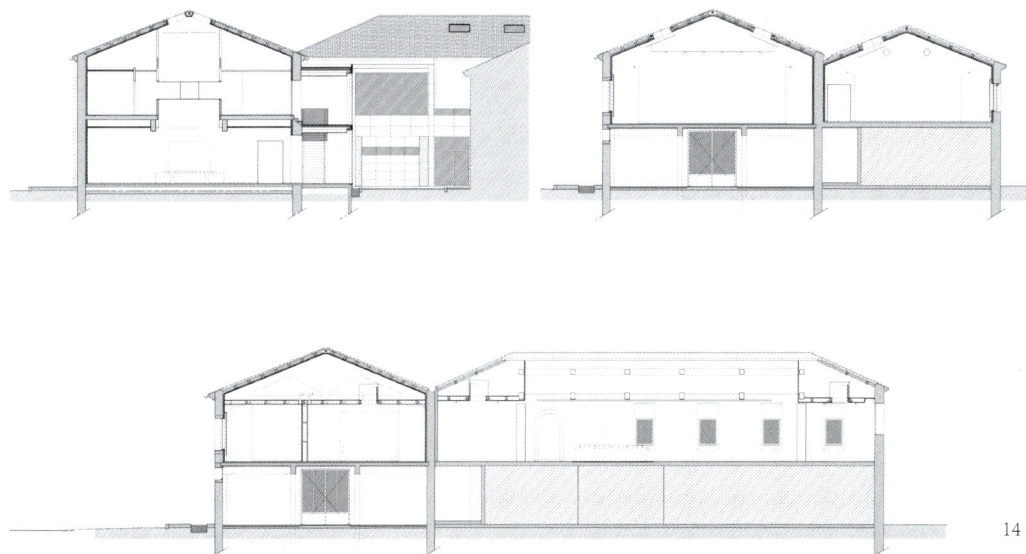

JEAN-FRÉDÉRIC OBERLIN MUSEUM

珍–弗莱德里克·奥柏林博物馆

建筑师 Architect: Frédéric Jung
地点 Location: Waldersbach
完工日期 Completion Date: 2003
摄影师 Photographer: Frédéric Jung
(pp.140-142), Michel Denancé
(pp.143-147)

The Jean-Frédéric Oberlin Museum is dedicated to the work of this 18th-century pastor and educationalist, a man open to the world who was both a witness of his time and an agent for change. Frédéric Jung undertook to convert or renovate the old buildings that date from different periods, from the presbytery (1789) and its common and the Froessel house (1724) to the Sophie Bernard extension (1978). This called for several different types of intervention, according to the intentions of

the museum, the problems encountered and the architectural project.

One of the major challenges was to connect the buildings to create a living museum presenting the collections and enabling people to "read" and question Pastor Oberlin's philosophy, through the creation of satellites organised around the presbytery, where the collection is housed. These satellites include the house of children and its educational workshops, the

temporary exhibition rooms, the archive and the thematic gardens.

The presbytery had to be preserved as a museum object in itself. Because there are exhibits from the 18th and the 20th century two forms of presentation and two approaches to the museum's functioning were necessary. They had to be clearly distinguished but linked. The presbytery and the Sophie Bernard extension are the results of these two approaches, one

synchronic, the other diachronic, one housing the collections in their place of origin, the other offering a contemporary reading of the collections – resonances in a consulting room open to the valley, and an observatory surveying the landscape of Ban de la Roche.

1. West façade
2. Museum extension
3. The observatory on the extension
4. The house of children
5. Master plan
6. Children's workshop

1. 西立面
2. 博物馆扩建
3. 扩建处的天文台
4. 儿童之家
5. 总体规划图
6. 儿童工作室

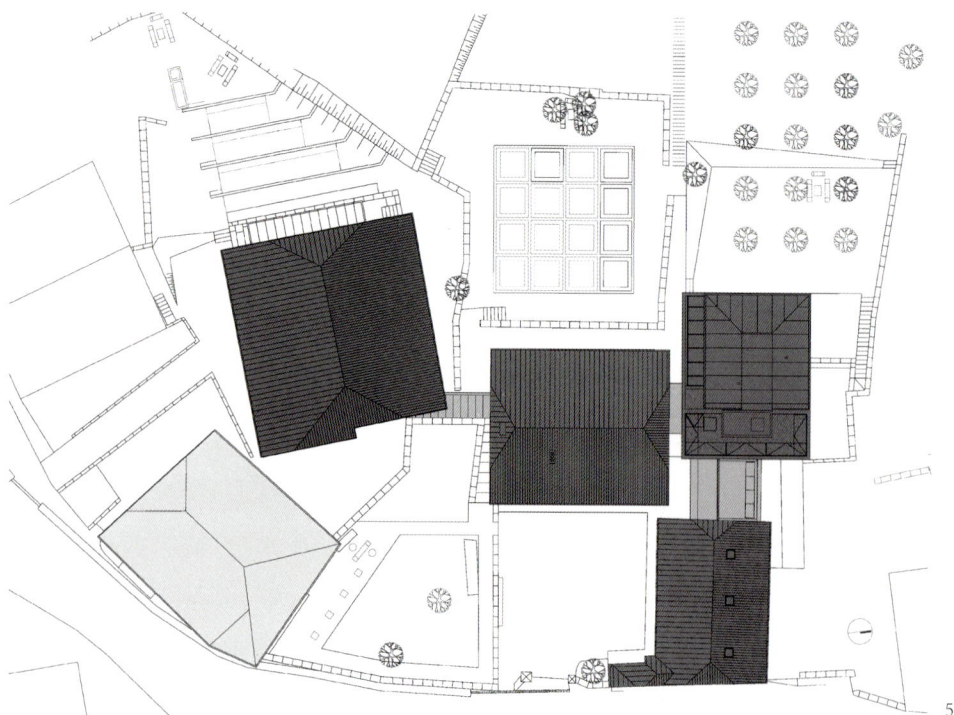

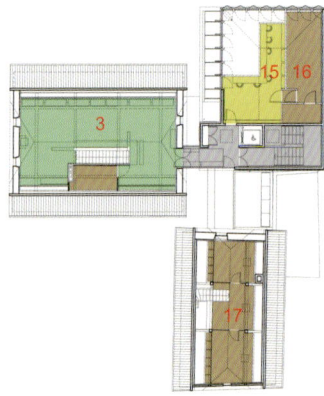

7

8

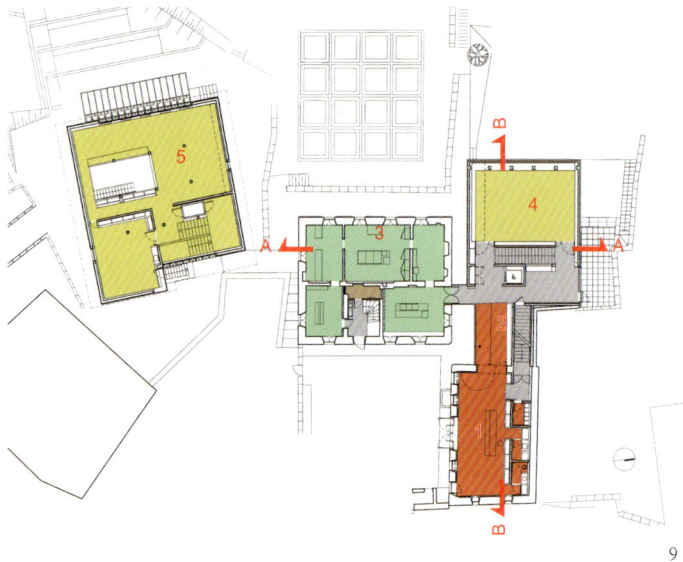

9

11

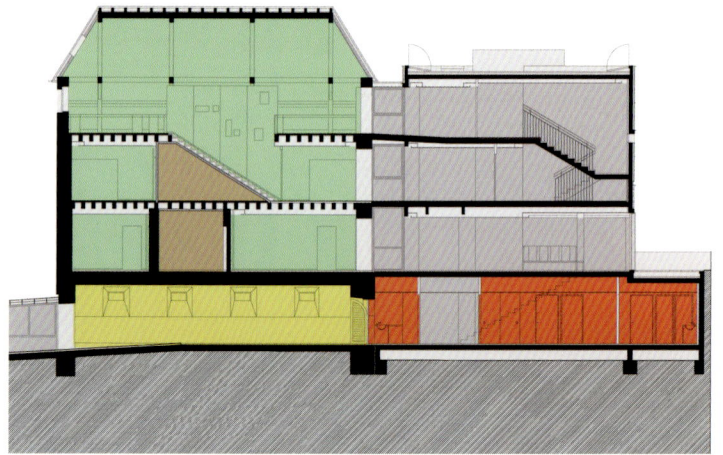

10

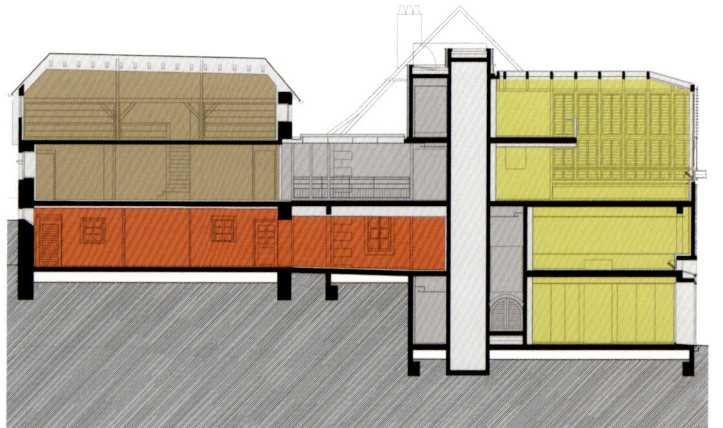

12

13

14

珍-弗莱德里克·奥柏林博物馆专为收藏18世纪伟大的牧师和教育学家奥柏林的作品而设计。奥柏林面向世界，既是自己时代的见证者，又是改革的代理人。弗莱德里克·荣格负责改造或翻新几座不同时代的旧建筑，从牧师宅邸（1789年）、弗洛塞尔住宅（1724年）到索菲·伯纳德的扩建工程（1978年）。根据博物馆的设计意图、所遇到的问题和建筑项目的不同，建筑师需要进行若干种不同类型的改造。

项目所面临的主要挑战就是将不同的建筑连接起来，形成一家鲜活的博物馆，让人们能够通过宅邸四周建筑的藏品解读并质疑奥柏林牧师的哲学。这些附属建筑包括儿童展厅、教育工坊、临时展厅、档案馆和主题花园。

牧师宅邸必须被保存成为博物馆。由于展览包含18世纪和20世纪两个年代，设计必须采用两种展示方式和两种功能设计方式。它们必须界限清晰又相互联系。牧师宅邸和索菲·伯纳德的扩建部分正好体现了这种诉求：一个限于一时，一个横跨历史；一个在原始地点收藏藏品，一个提供藏品的现代解读——二者在一间朝向山谷的咨询室汇合，从瞭望台上可以观望拉洛查地区的景色。

15. The museum collections
16. Link between the presbytery and the extension
17. Temporary exhibition room
18. Reading room

15. 博物馆收藏
16. 牧师宅邸和扩建部分之间的连接
17. 临时展区
18. 阅读室

16

17

18

LILLE MODERN ART MUSEUM – LAM

里尔现代艺术博物馆

建筑师 Architect: Manuelle Gautrand
地点 Location: Lille
完工日期 Completion Date: 2010
摄影师 Photographer: Max Lerouge

The original museum, designed by Roland Simounet and opened in 1983, was a horizontal structure stretching out lengthways in a lightly undulating park. The west wing housed the administrative offices and the east wing the auditorium and the exhibition rooms. The central section housed the entrance – a glazed gallery which gave onto an interior patio, offering the main breathing space of the museum. In 2000 the building was listed as a Historic Monument.

The donation of a collection of more than 3000 works of Art Brut prompted the museum to embark on a modernisation and extension project. The competition was won by Manuelle Gautrand. Two complex and sensitive challenges are brought together in this project: to extend the work of Roland Simounet, and to invent a place to house this magnificent collection. Manuelle Gautrand did not want to create an independent additional sequence, but rather a connection. The project is built from the starting point of a

large, gentle form, extending from the west wing of the site to the east. From the south side, visitors discover the new extension at the end of their visit, behind the existing sequences.

The fold in the ground meant that the central portion of the structure had to be narrow. From both sides of this pintucked centre, the volume becomes suppler and comes loose in the form of two fantails enveloping the existing museum. One, in the west, is designated for functions other

3

4

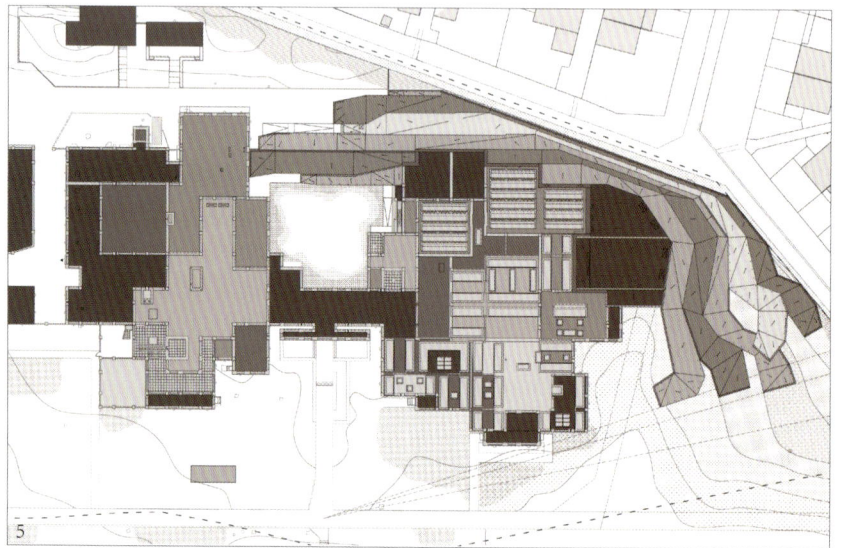

5

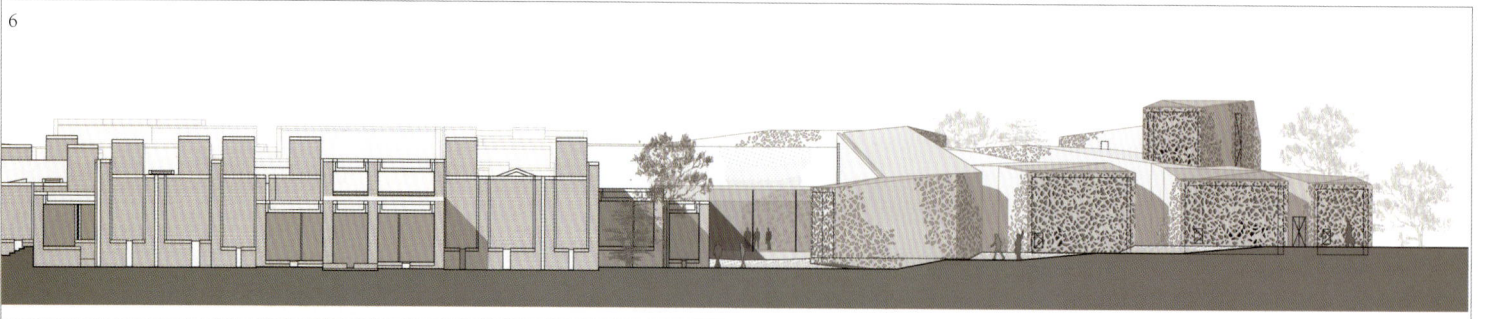

6

than exhibiting (a technical space for the works, cafeteria, etc.); the other, in the south-east, is devoted to exhibiting the Art Brut collection. These fans are made up of five folds, following the lay of the land, each one housing a different theme from the collection. Where the original architecture is orthogonal, strict and ordered, that of the extension becomes more supple, more organic, expressing the "envelopment" necessary for presenting Art Brut.

In the interior modifications carried out for the existing museum, the reception remains at the geographical centre of the project, leading to the exhibition spaces, the bookshop that runs along the patio, the

cafeteria that connects to the auditorium. The storerooms and restoration workshops have been enlarged and modernised. The exhibition spaces have been entirely restructured. In order to bring the security up to date, three almost invisible compartments have been created, while respecting the original architecture. Finally, for the Art Brut spaces, a glazed gallery allowed the architect to extend the original east-west museographic itinerary and prepare the visitor to enter the first room. This forms a fulcrum, opening onto the five Art Brut rooms, which are supple forms that slide into the landscape. Each opening onto the exterior has been designed to create a visual detour, punctuating the visit with escapes into fantasy.

10

8. The pivot room, looking towards the existing museum
9. Inside the extension, at the extremity of a "finger"
10. The extension – interior view

8. 位于新旧博物馆之间的枢纽地带，通往原有博物馆区域
9. "手指"状扩建空间的末端
10. 扩建空间的内景

原有的博物馆由罗兰德·西莫尼特设计，于1983年对外开放，是一个建在地势起伏的公园里的纵向延伸的水平结构。建筑西翼是行政办公室，东翼是礼堂和展览厅，中央部分是入口。玻璃走廊直通室内天井——博物馆的主要休息空间。2000年，建筑被列入历史纪念物名录。

超过3,000件原生艺术作品的捐赠让博物馆登上了现代艺术的舞台并促使其兴建扩建项目。最终，设计竞赛由曼纽艾乐·高童获得优胜。项目面临着两个复杂而敏感的挑战：一是对罗兰德·西莫尼特的建筑作品进行扩建，二是创造一个全新的空间来放置藏品。曼纽艾乐·高童并不想建造一个独立的附加结构，而是想要一个连接物。项目从原有博物馆西翼延伸至场地东侧。当参观旅程结束时，参观者能够从南面发现原有建筑后面的扩建结构体。

折叠结构意味着扩建部分的中央部分必须是狭窄的。这个折叠中心的两侧，空间呈扇尾状逐渐变得宽敞，包裹住原

有的博物馆。西面的部分设置着除展览以外的功能区（工作技术区、餐厅等），东南部分则用于展示原生艺术藏品。扇形结构分为五折，随着地势而建，各自拥有不同的展览主题。原有建筑四四方方、精密而有序，而扩建结构则灵活而富于变化，反映了原生艺术的特色。

在原有博物馆的室内改造中，前台保持了地理中心的位置。参观者由前台进入展览空间，沿着天井的树顶，连接礼堂的餐厅。库房和修复工作室经过了扩建和现代化改造。整个展览空间都经过了重新构建。为了更新安保设施，在尊重原始建筑的基础上，建筑师创造了三个几乎无形的隔间。最后，在原生艺术区，玻璃走廊让建筑师得以将原来东西向的展览路线延伸，让参观者做好准备进入第一个展厅。以此为支点，五个原生艺术展厅依次展开，形成了与景观相辅相成的灵活空间。每个朝向室外的开口都形成一个视觉通道，让人们富于想象。

11. Purple model of the extension
12. Cross section
13. Ground floor plan
14. The extension – interior view

11. 紫色的扩展建筑模型
12. 横截面
13. 一层平面图
14. 扩建部分内景

1. Entrance hall
2. Patio
3. Administrative offices
4. Auditorium
5. Exhibition rooms
6. Functional spaces (technical space, cafeteria...)
7. Exhibition rooms for the Art Brut collection

1. 入口大厅
2. 中庭
3. 行政空间
4. 演映厅
5. 展览空间
6. 功能性空间
7. 原生艺术展览空间

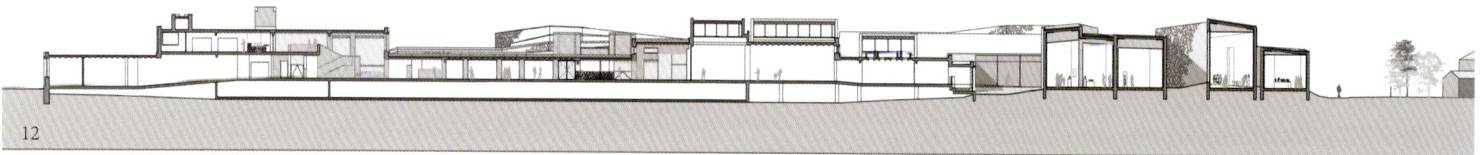

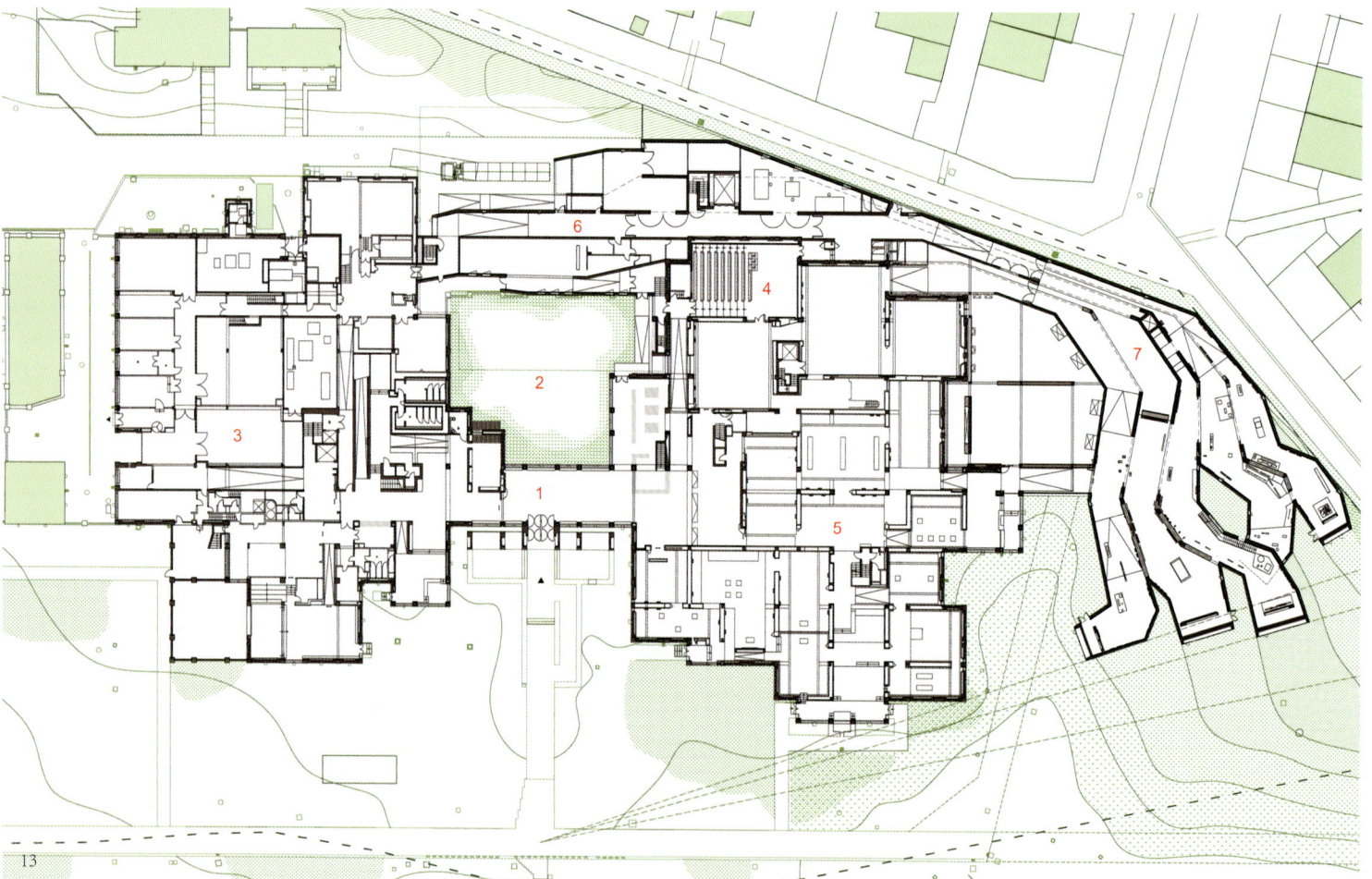

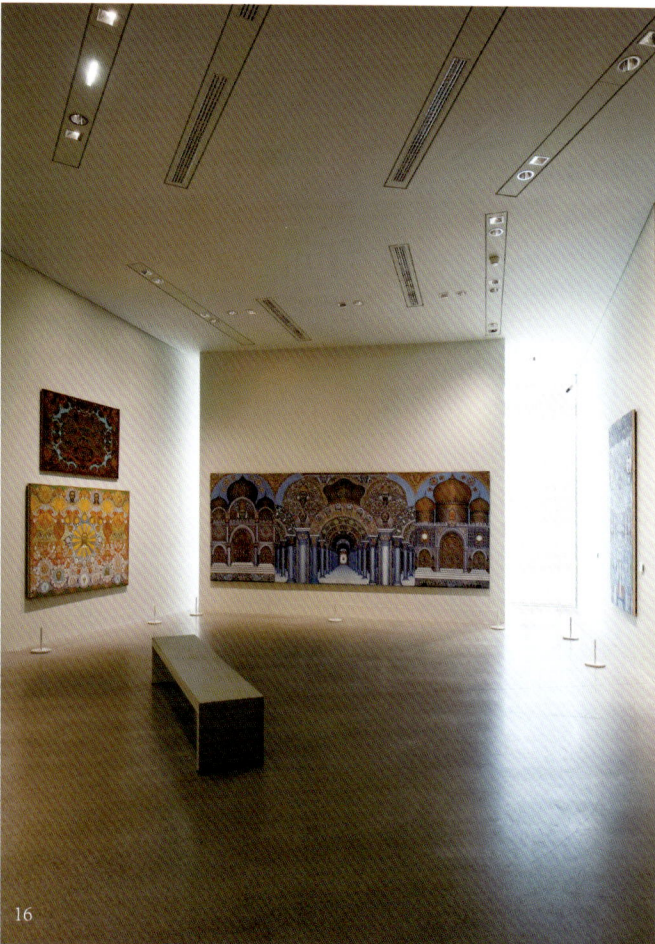

15-17. The extension – interior view
18. Interior view of the extension's "moucharaby"

15-17. 扩建部分内景
18. 扩建部分的穆沙拉比窗的内景

NANCY MUSEUM OF FINE ARTS

南希美术馆

建筑师 Architect: Beaudouin-Husson
地点 Location: Nancy
完工日期 Completion Date: 1999
摄影师 Photographer: Luc Boegly (pp.158-
159), Jean-Marie Monthiers (pp.160-163),
Olivier-Henri Dancy (p.165)

The extension of the Museum of Fine Arts on Place Stanislas is an abstract work on gravity, light and time. It is composed of three rectangular planes that cannot be seen simultaneously and can only be understood over the course of an exploration of the museum. The architecture calls on our memory of the space.

These three planes have different relationships with gravity. The first is in dark granite placed on the ground; it slides under a second plane of paler stone which appears to be suspended horizontally. The third is white, floating freely in the interior space of the museum. The essence of the project is found in these three surfaces, which establish a metaphorical relationship between the sky and the earth. The museum space moves freely between the exterior and the interior, surrounded by these three elements.

The building is anchored in the history of the site. It is low and horizontal so as not to

upset the volumetry of the square. It is like a weight that nevertheless removes the effect of gravity, raised above a transparent space, like its model, the Palais de l'Intendance, a mature masterwork of the architect Emmanuel Héré, who, by the 18th century, had already abandoned formal conventions in favour of more abstract intentions: horizontality, lightness, transparency.

1. View of the museum from the garden
2. Staircase built in 1936
3. Garden side façade
4. The ramp
5. Master plan

1. 临花园的博物馆
2. 1936年建成的楼梯
3. 临公园的博物馆立面
4. 坡道
5. 总体规划图

斯坦尼斯拉斯广场上美术馆的扩建是一项针对重力、光线和时间而进行的抽象工作的呈现。它由三个不能同时看见的矩形平面组成，只有在走过参观路线之后，人们才能理解它的空间结构。建筑借此召唤着人们的空间记忆。

三个平面与地心引力有着不同的关系。第一个平面采用了黑色花岗岩铺地，隐藏在以灰白石头铺地、看似水平悬浮的第二个平面之下；第三层平面则为白色，自由地飘浮于博物馆的室内空间之中。项目的精髓就在于这三层暗喻着天空与大地的平面。博物馆空间徜徉在这三个元素之间，从室外到室内自由地游走。

建筑扎根于场地的历史。低矮的水平造型使得广场能够维持适当的体量比例。博物馆就像脱离了重力影响一样，坐落于一个透明空间的上方，正如它的建筑原型——总督厅。总督厅是建筑师艾曼纽·海莱的一件名作，他在18世纪就已经摒弃了传统的形式惯例，主张更抽象的设计：水平、轻盈、通透。

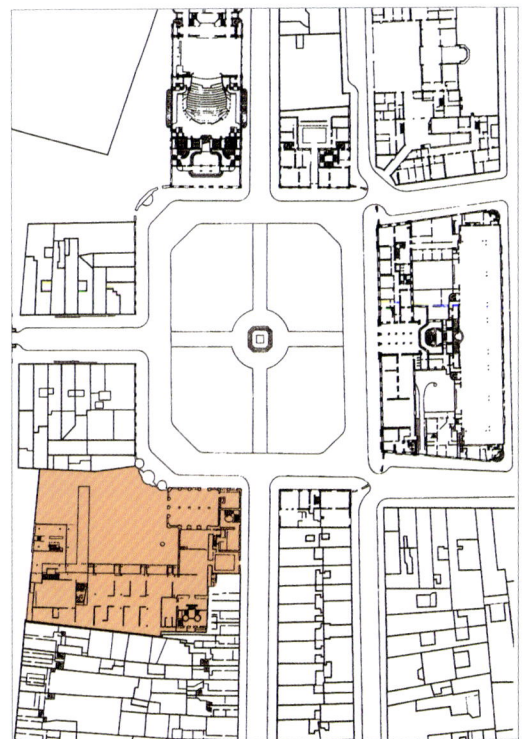

6. Garden floor
7. Ramp leading to the 1st floor
8. Gallery
9. Longitudinal section

6. 花园一层
7. 通向二层的坡道
8. 长廊一景
9. 纵剖面

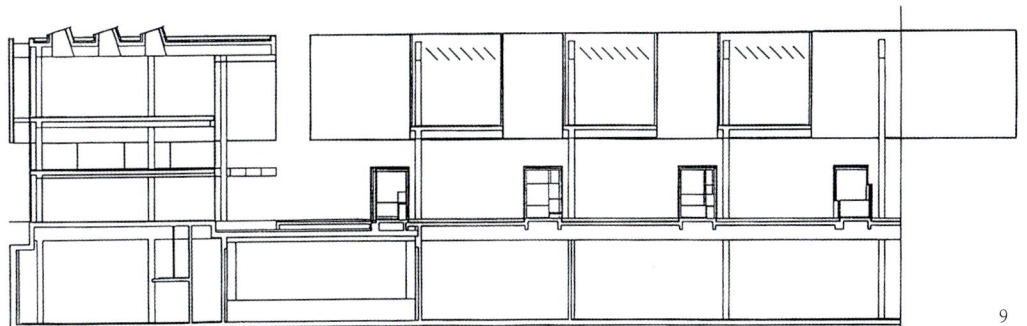

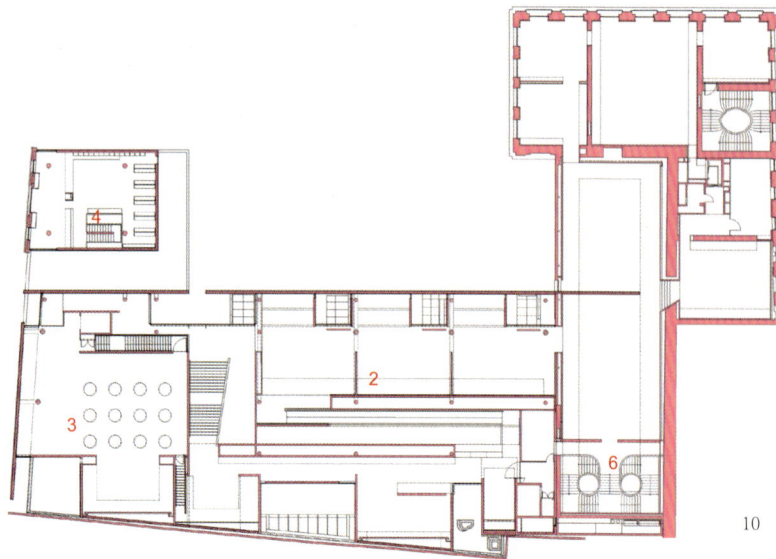

10

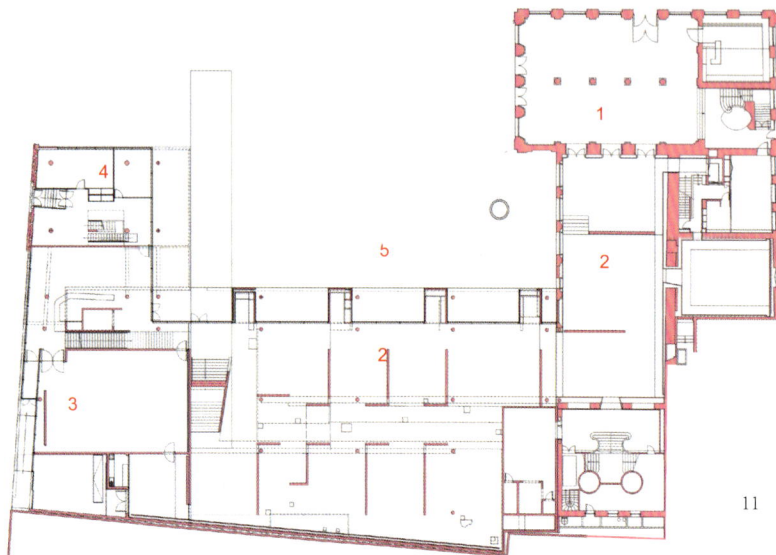

11

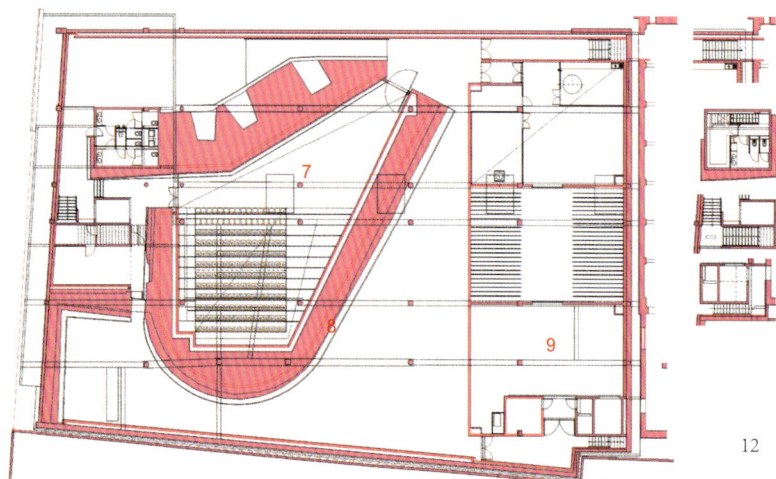

12

13

14

INTERNATIONAL CITY OF
LACE AND FASHION

国际蕾丝时装城

建筑师 Architect: Moatti-Rivière
地点 Location: Calais
完工日期 Completion Date: 2009
摄影师 Photographer: Michel Denancé
(pp.166-167), Florian Kleinfenn (pp.168-169),
Moatti-Rivière (pp.170-175)

The City extends over 7,500m² and is made up of two main buildings. The first, in the old Boulart factory that dates back to 1870, contains the permanent collections and the administrative offices. The second is the extension (2500m²) for temporary exhibitions. An auditorium seating 200 links the two. The old lace factory is made up of two main blocks with four floors, linked to form a U shape with a large courtyard in the middle. The metallic-framed extension has a Z shape, and forms a rigid box on five faces (roof, floor and three façades – north, south and east) with the sixth open in a double-curve façade. The steel structure allowed for a 17m cantilever.

Moatti et Rivière have designed an usual double-skin façade in tempered glass over the continuous curve of the façade. It is heat-curved and screen-printed on the outer face. The screen-printing

complicated the operation: the motifs had to be adapted to avoid geometrical distortions. The precision of the work and the size of the panels (1.6m x 1.6m) meant they had to be placed and joined with great care. The panels are connected by cleats in moulded stainless-steel with a brushed finish. The printed motif represents the stitches of the Jacquard cards from the Leavers enterprise. A mechanical ventilation between the two skins insulates

the building, avoids condensation and allows for the surfaces to be cleaned.

The City is entered through the extension, offering a new face to the old factory. The path of the permanent exhibition traces the industrial history of lace in Calais, through five sequences unfolding in rooms distributed over three of the four levels of the old factory. The main staircase, in the centre of the building, has been used to link them scenographically. The third floor is reserved for the administrative offices. The complete gutting and restructuring the Boulart factory has enabled it to welcome the public with a museography freed of all constraints.

1. The new façade
2. The façade at night
3&6. The old factory
4. Master plan
5. Elevation of the façade

1. 新立面
2. 立面夜景
3&6. 旧工厂外观
4. 总体规划图
5. 立面图

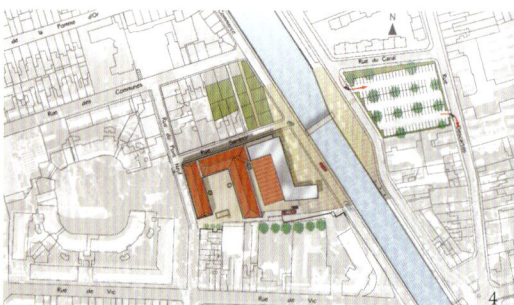

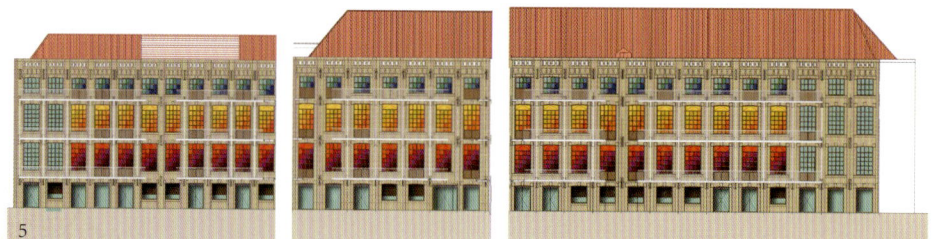

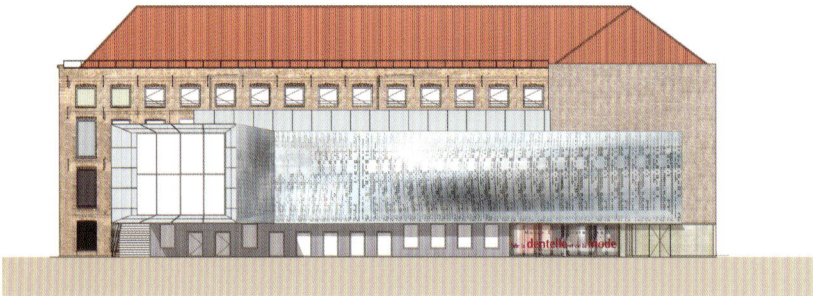

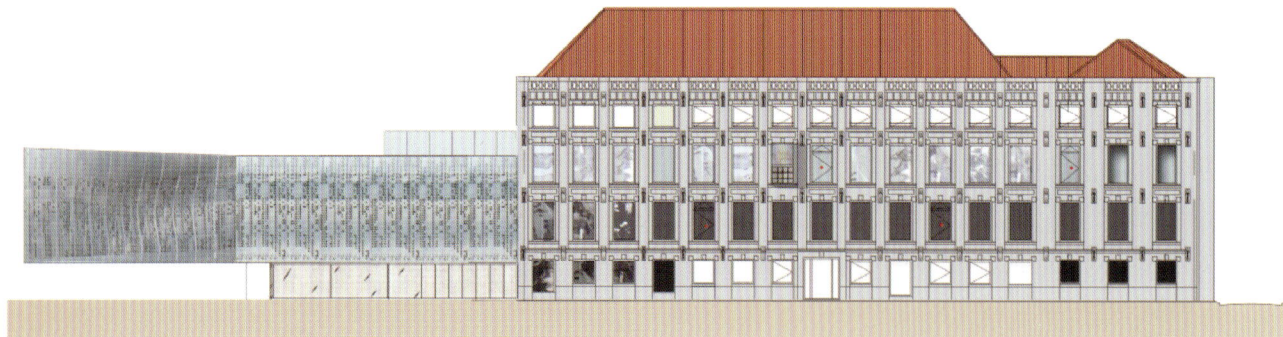

项目面积7,500多平方米，有两座主楼组成。第一座是1870年建成的布拉特工厂，里面设置着永久藏品和行政办公室。第二座是扩建结构（2,500平方米），用于临时展览。一座可容纳200人的礼堂将二者连接起来。旧蕾丝工厂由两个主体组成，分为四层，由中间的庭院连接成U形。金属结构的扩建部分呈Z形，形成了一个五面的盒子（屋顶、楼面和三个外立面——北、南、东），第六面采用了双弧度外立面。钢铁结构由一个17米的悬臂构成。

在弧形外立面上，Moatti et Rivière设计了一个普通的双层钢化玻璃立面。它经过高温压弧，并在外立面上进行了丝网印刷。丝网印刷让操作过程复杂化：图案必须避免几何变形。工作的精确度和面板的尺寸（1.6米X1.6米）意味着必须精心制作。面板通过经刷面处理的不锈钢夹板相连。印花图案呈现了列韦斯公司的提花纹板针法。两层表皮之间的机械通风起到了隔热作用，避免了水汽凝结，并且保证了表面的洁净。

蕾丝时装城由扩建部分进入，为旧工厂提供了一个全新的形象。永久性展区的路线沿着加来的蕾丝制造历史展开，通过分散在旧工程三层楼（工厂共有四层）之间的五个序列来呈现。建筑中央的主楼梯将它们连接起来。第四层被保留用作行政办公空间。布拉特工厂的整体改造和重建使其能够以不受限制的展览空间和场景布置开放给公众参观。

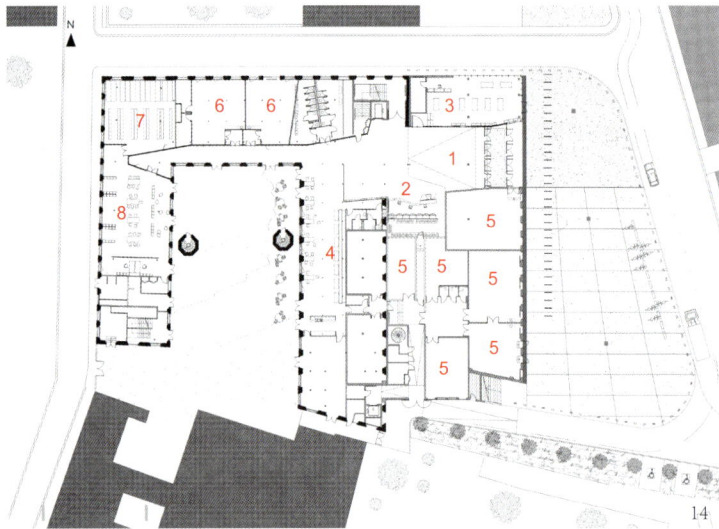

13. Museography, part 1, the origin of lace
14. Ground floor plan
15. Museography, part 2, lace and fashion
16. 1st floor plan
17. 2nd floor plan

13. 博物馆展示，第1区，蕾丝的起源
14. 一层平面图
15. 博物馆展示，第2区，蕾丝与时尚
16. 二层平面图
17. 三层平面图

1. 入口大厅	1. Entrance hall
2. 接待处	2. Reception area
3. 商店	3. Shop
4. 餐厅	4. Restaurant
5. 维修工坊	5. Restoration workshop
6. 教学空间	6. Educational space
7. 资料查询处	7. Documentation
8. 图书馆	8. Library
9. 永久性展览室入口	9. Permanent exhibition entrance
10. 花园展示	10. Garden
11. 工厂展示	11. Factory
12. 当代蕾丝展示	12. Lace today
13. 大厅	13. Hall
14. 演映厅	14. Auditorium
15. 临时性展览室	15. Temporary exhibition room
16. 工作室	16. Workshop
17. 镜厅	17. Mirrors gallery
18. 临时性展览室上方挑空	18. Atrium overlooking the temporary exhibitions
19. 大厅上方挑空	19. Atrium to the hall

15

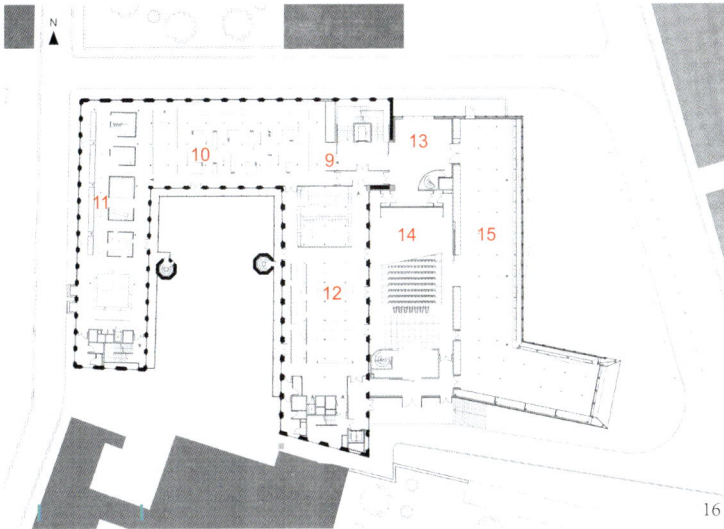

16

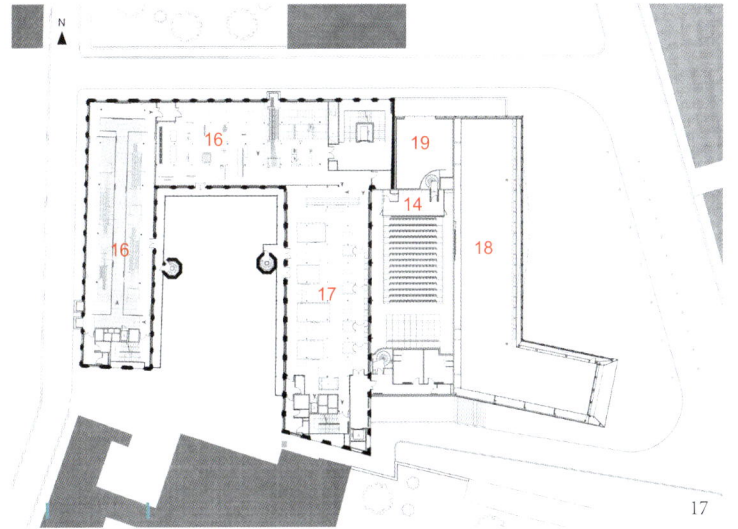

17

18

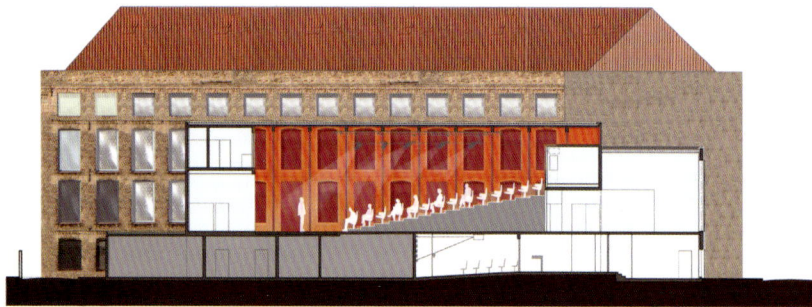

19

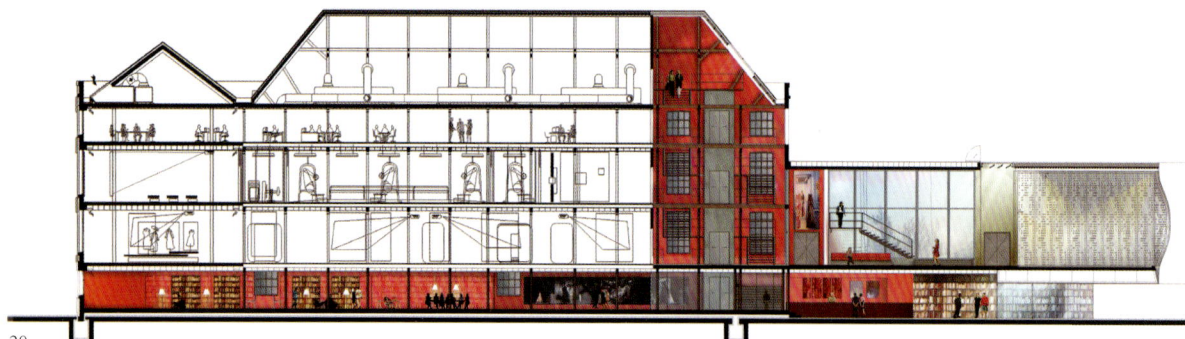

20

18. Museography, part 2, lace and fashion
19. Section of the auditorium
20. Cross section
21. Interior view, detail

18. 博物馆展示，第2区，蕾丝与时尚
19. 礼堂剖面图
20. 横截面图
21. 内景

PRESIDENT JACQUES CHIRAC MUSEUM

希拉克总统博物馆

建筑师 Architect: Jean-Michel Wilmotte
地点 Location: Sarran
完工日期 Completion Date: 2006
摄影师 Photographer: Fabre (pp.176-178), Gratien (pp.179-180), MPJC (p.181)

In 1998 Jean-Michel Wilmotte won the competition to design the President Jacques Chirac Museum. The construction's insertion into the village of Sarran and the use of the traditional materials of Corrèze architecture ensured a perfect integration of the museum into a protected environment. Two barns formed the departure point for the project. Now restored, they house the restaurant and a small storage locale.

The initial museum was inspired by the proportions and the long volumetry of these two pre-existing buildings, which are typical of the vernacular architecture of the Corrèze. The architects added two buildings, one for the permanent exhibition, the other for temporary exhibitions, covered with a double-slope slate roof. They are linked by a covered gallery that forms the hall of the museum. Granite, slate, oak and chestnut, as well as its division into several units, give it the appearance of

a small hamlet and aid the integration of the museum into its environment; steel, concrete and glass anchor it in modernity. The public spaces are vast, open to each other and benefit from direct or indirect natural light in a warm ambiance.

In 2006, the spirit of a contemporary hamlet was completed by two buildings that appear independent but are in fact linked below ground. The first building, which has four floors including two below

5

1. General view of the museum
2. Footbridge from the library and the esplanade to the museum
3. The museum at night
4. The entrance
5. The museum seen from the park

1. 博物馆外景
2. 从图书馆与观景平台通往博物馆的行人桥
3. 夜晚时的博物馆
4. 入口处
5. 临公园博物馆一角

ground, houses the administrative offices, cultural activities and the restoration and preservation of the collections; the second building, a five-floor tower, houses the library, a space for exhibiting files and storerooms. The tower and the administrative building are linked by a covered footbridge. A wide esplanade, under which are found a large storage reserve that can be visited and an auditorium, link the old building to the new. The architectural approach is homogenous, inspired by the first museum and showing care to integrate into the site through the alternate use of granite, concrete, slate, glass and wood.

9

6. Reflection of the museum restaurant
on the north glass façade
7. Interior of the library
8. Introduction to the museum's
permanent exhibition
9. Temporary exhibition room

6. 北面镜墙所反射出的博物馆餐厅
7. 图书馆内部
8. 博物馆常设展览厅入口
9. 临时展览室

1998年，让-米歇尔·威尔莫蒂获得了希拉克总统博物馆设计竞赛的优胜。博物馆坐落在萨兰村内，采用了科雷兹省建筑的传统材料，与受保护的环境完美地结合起来。两座谷仓形成了项目的出发点。经过修复，它们分别是餐厅和小型仓库。

博物馆的设计从这两座原有结构的比例和细长造型（科雷兹省的典型建筑）中获得了灵感。建筑师添加了两座建筑，一座用于永久性展览，一座用于临时展览。二者采用了相同双坡板岩屋顶，通过作为博物馆大厅的带顶走廊连接起来。花岗岩、板岩、橡木和栗子木以及建筑的单元分割方式让它看起来像一个小村庄，使博物馆进一步融入了环境之中；钢铁、混凝土和玻璃则赋予博物

馆现代特征。公共空间十分宽敞，相与连接。得益于直接、间接的自然光，沐浴在温暖的氛围中。

2006年，两座看似独立、实则于地下相连的建筑完善了现代村庄的设计。第一座建筑共有四层（其中地下两层），内设行政办公区、文化活动区和藏品修复保护区；第二座塔楼高五层，内设图书馆、文档展览区和储藏室。塔楼和行政楼通过带顶的人行桥相连。宽敞的大道下方是可供参观的大型仓库和礼堂，将新旧建筑连接起来。建筑设计具有一致性，与第一座博物馆都采用了花岗岩、混凝土、岩板、玻璃和木材。

BRAND NEW
EXPRESSIONS
全新的发挥

POMPIDOU CENTRE
蓬皮杜中心

建筑师 Architect: Renzo Piano, Richard Rogers
地点 Location: Paris
完工日期 Completion Date: 1977
摄影师 Photographer: Renzo Piano Building Workshop

Occupying an entire block and covering approximately 100,000m^2 at the heart of Paris, the Pompidou Centre is devoted to the plastic arts, music and reading. Its architecture, based on a large rectangle at its base, with all its practical functions (air conditioning, water, mobility systems) on the outside, symbolises contact between surfaces or between people indoors or out. Like the Eiffel Tower in its day, the Pompidou Centre sparked passionate debate among the public and politicians at every stage of its design and construction.

The construction work posed numerous technical and logistical problems that were resolved only by great conviction and determination. At the opening, in 1977, the public immediately embraced the Pompidou Centre, its museum of modern art, the square and its neighbourhood. Students and researchers readily adopted its library, and tourists gladly took the trip up its winding, caterpillar-like exterior glass escalator to gain unparalleled views of the neighbourhood and the city.

After 20 years of operation, the museum needed more space, and the library had to be made more accessible. The first major decision of the overhaul was to relocate all of the offices to a neighbouring building. Spaces were thus freed up for the centre's primary activities. The fourth floor houses the museum of modern art, and the fifth is reserved for temporary exhibitions. The library, which is located on the first, second and third floors, has a separate entrance for ease of access. The main ground-floor forum was also modified: a large opening in

the sweep of the main entrance connects the three underground levels (performance rooms), the ground level (entrance, ticket office and information desk) and the two mezzanine levels (boutiques and café). All are linked by elevators and stairways.

Aside from cleaning, the only change to the exterior was the introduction of a weather-proof canopy over the main entrance.

1. The building after restoration, night view
2-3. Details of the façade
4. The terraces of the contemporary art museum
5. General view
6. View from the escalator

1. 翻修后的建筑夜景
2、3. 立面细部
4. 当代艺术博物馆的天台
5. 外观全景
6. 电梯一景

蓬皮杜中心在巴黎的中心占据了整个街区，占地面积约100,000平方米，致力于传播造型艺术、音乐和阅读。它的建筑结构以巨大的长方形底座为基础，所有实用功能（空调、水、移动系统）都暴露在外，象征着表面和室内外的人们之间的联系。就像埃菲尔铁塔一样，蓬皮杜中心在设计和建造的各个阶段都激发了公众和政界人士之间的激烈讨论。

建造工作面临了无数的技术与运筹问题，只有通过巨大的决心和信念才能解决。1977年对外开放以后，公众立刻接受了蓬皮杜中心，包括它的现代艺术博物馆、广场和周边建筑。学生和研究者都乐于到图书馆进行研究，而游客则喜欢沿着它蜿蜒的、如毛毛虫一般的室外玻璃扶梯领略周边和城市的风光。

在运营20年之后，博物馆需要更多的空间，而图书馆也需要更容易进出。管理人员决定将所有办公设施都移到旁边的大楼中，空出的空间正好用以进行中心的主要活动。五楼设置着现代艺术博物馆，六楼则用于临时展览。图书馆位于二到四楼，设有独立的入口。一楼的论坛也经过了改造：横跨主入口的开口连接了地下三层空间（表演厅）、一楼（入口、售票处和信息台）以及两个中层楼（精品店和咖啡厅）。所有空间都由电梯和楼梯相连。除了清洁之外，外立面唯一变动在于主入口处引入了一个防风雨的华盖。

7&9. The new main lobby, the focal point
of all activities and events
8. Cross section
10. Main façade (west) and section of the
forecourt with an underground car park

7、9. 新修大堂，所有活动的焦点空间
8. 横剖面
10. 西侧主立面和前广场（带有地下停车场）
剖面图

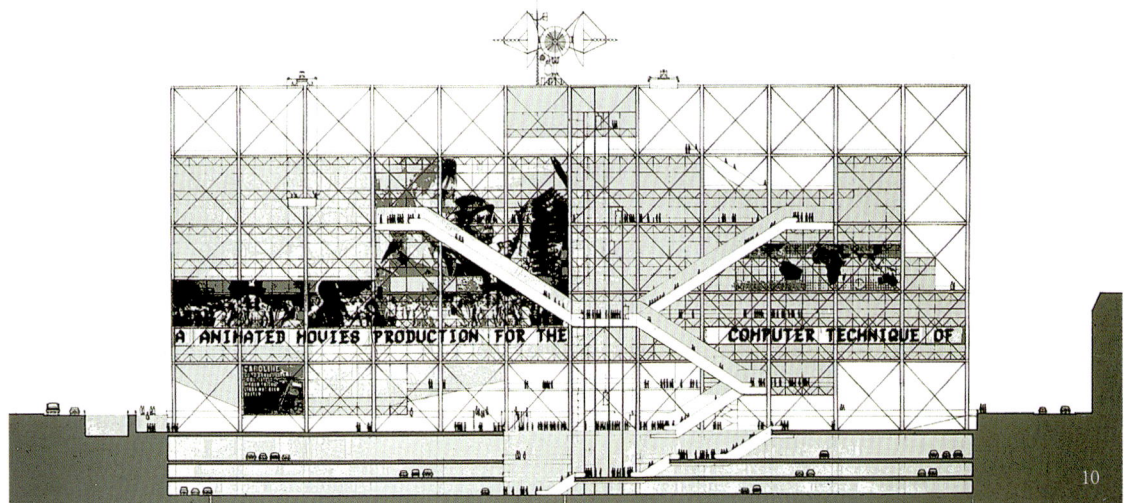

ÉRIC TABARLY CITY OF SAILING

埃里克·塔巴里航海城

建筑师 Architect: Jacques Ferrier
地点 Location: Lorient
完工日期 Completion Date: 2007
摄影师 Photograph: © jfa / Jacques Ferrier
Architectures – photo Luc Boegly (pp.190-
191, 193-195), photo Jean-Marie Monthiers
(p.192)

A symbol of the revival of the submarine base of Lorient, the Éric Tabarly City of Sailing, covering 6,700m² with 2,500m² devoted to exhibitions, has sprung up in the heart of the old military compound. In contrast to the grounded solidity of the three concrete bunkers, the project offers a shiny metal vessel hanging over the quay. Floating above a transparent ground floor, this sparkling craft is moored to the "tower of the winds", a vertical signal tower placed on a sea arm, around which pontoons are moored.

Covered in iridescent aluminium panels that change colour with the weather, the vessel hangs over the quay to form a spectacular canopy facing the sea. The glazed façade, sheltered by the cantiliver, reveals the activity inside. The reception with its uneven geometry serves as an entrance to the exhibitions and to the auditorium, as well as to the restaurant. The exhibition hangar, accessible from the ground floor, rises in double height to the first floor, with an exhibition room and a space for the maintenance and repair

of boats. The first floor is a wide platform open to the sea, around which exhibition spaces devoted to sailing and yachting are organised. This platform is covered by a hull whose interior surfaces are lined in wood, in reference to naval architecture. A long footbridge extends from the south façade of the building and links the exhibition level with the tower of the winds viewing platform and with the pontoons.

Cap l'Orient is leading the way in sustainable and community-based development, as the

Éric Tabarly City of Sailing shows. The building was designed according High Environmental Quality (HQE) standards, to reduce its energy consumption as much as possible. Its well-conceived architecture allows it to use solar energy and the natural resources of the site, while its technology ensures low energy use.

1. Entrance
2. Exhibition hangar
3. Repair hangar
4. Boarding lounge
5. Shop
6. Auditorium
7. Sailing News area
8. Bar and storage
9. Restaurant
10. Kitchen
11. Mixed public restrooms

1. 入口
2. 展览棚
3. 维修机库
4. 登船休息室
5. 商店
6. 礼堂
7. 航海新闻区
8. 酒吧和储藏室
9. 餐厅
10. 厨房
11. 公共洗手间

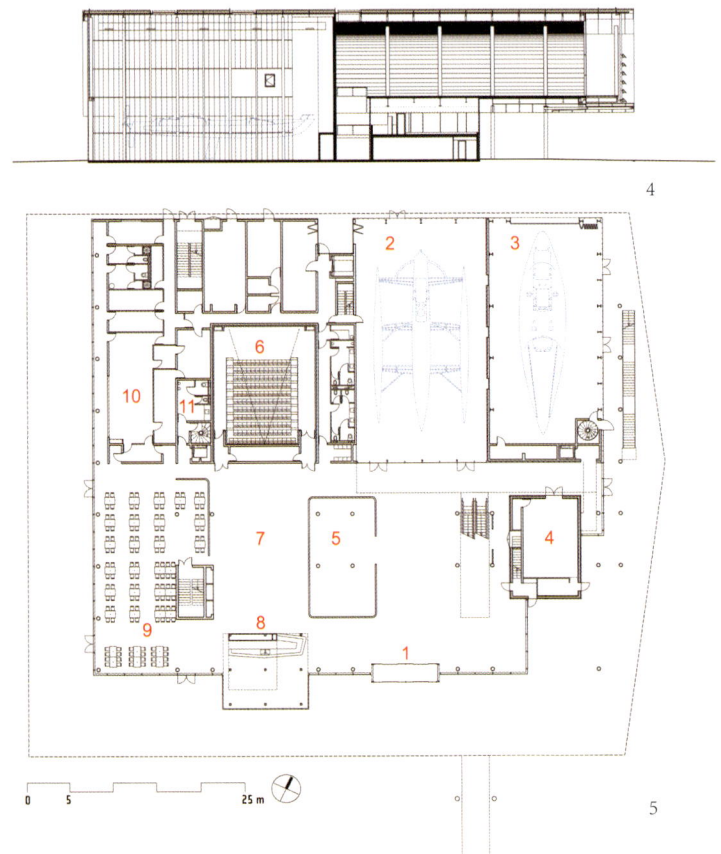

1. The City, anchored in the water
2. The south façade, covered in photovoltaic panels
3. The Tower of the Winds signal tower
4. Longitudinal section
5. Ground floor plan
6. A boarding quay to get a first taste of sailing
7. Symbol of the renewal of the submarine base of Lorient

1. 锚定在水中的航海城
2. 覆盖着太阳能光伏板的南立面
3. "风之塔"信号台
4. 纵剖面
5. 一层平面图
6. 帆船码头
7. 洛里昂潜艇基地重建的象征

作为洛里昂潜艇基地复兴的象征，总面积6,700平方米（其中2,500平方米专门用于展览）的埃里克·塔巴里航海城在旧军事基地上拔地而起。项目与三座混凝土贮仓坚固的形态不同，以闪亮的金属船舰形象伫立在码头。悬浮在透明的底楼上，闪闪发光的建筑停泊在"风之塔"旁。风之塔是海湾上一座垂直的信号塔，四周停泊着浮码头。

建筑外观的闪光铝板会随着天气的变化而变换颜色，在码头呈现出一道壮丽的华盖。悬臂下方的玻璃外墙展示出内部的活动。不规则几何造型的前台是展览厅、礼堂和餐厅的入口。从一楼进入的展览区有两层楼高，分为展览厅和船舶养护区两部分。二楼宽敞的平台朝向大海，四周环绕着帆船和游艇展览空间。平台外壳的内表面由木条组成，与大船的建造有异曲同工之妙。从建筑南侧延伸出来的长桥将展览楼层与风之塔的观景平台和浮码头连接了起来。

埃里克·塔巴里航海城项目体现了洛里昂在可持续和本位发展方面的领军地位。建筑的设计严格遵照了高环境品质标准，尽量减少了能源消耗。精心构思的建筑利用了太阳能和当地的自然资源，而现今的技术则保证了低能源消耗。

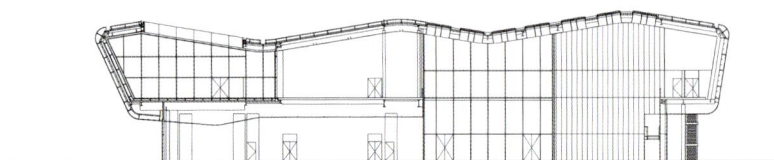

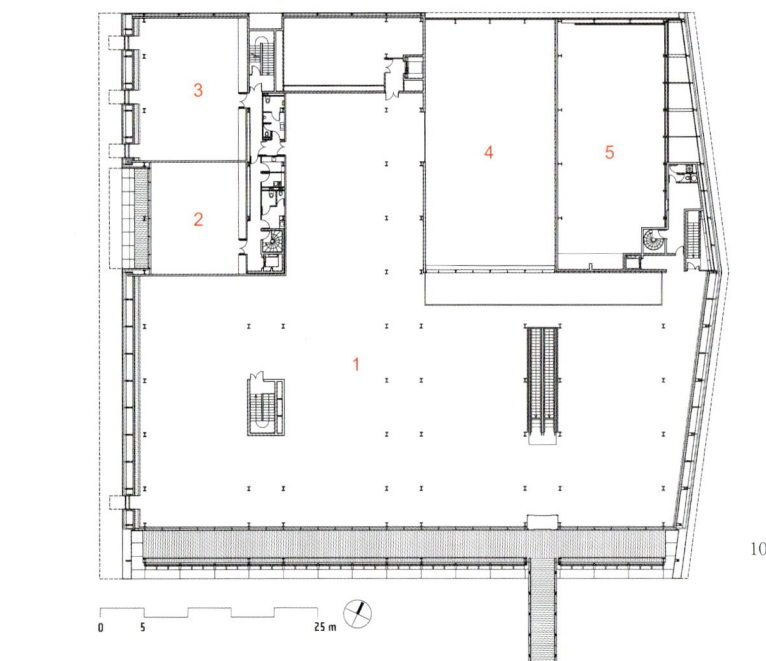

8. The first floor, where exhibitions are organised
9. Cross section
10. 1st floor plan
11. The loggia offers a view over the bay

8. 二层展览区平面图
9. 横截面
10. 二层平面图
11. 可以俯瞰海湾的长廊

1. Exhibition room
2. Library
3. Offices
4. Atrium over the exhibition hangar
5. Atrium over the repair hangar
6. Crew room

1. 展览室
2. 图书馆
3. 办公室
4. 展览棚上方的挑空
5. 维修机库上方的挑空
6. 船员室

1

CITY OF THE OCEAN

海洋城

建筑师 Architect: Steven Holl &
Solange Fabião
地点 Location: Biarritz
完工日期 Completion Date: 2011
摄影师 Photographer: Laurent Garcia
(pp.197 top, 198-205), Emmy Martens
(p.196), Balloïde Photo (p.197 bottom)

Designed by the New York architect Steven Holl and his Brazilian colleague Solange Fabião, the City of the Ocean espouses the form of the waves on the Basque coast. Its original architecture, dreamed up by this international duo, is based on the spatial concept "Beneath the sky, beneath the ocean", with a geometry of convex and concave surfaces.

The main building houses a science-through-play area, exhibitions and an auditorium as well as offices. A restaurant with a terrace, a cafeteria and a snack bar that can be entered separately from the exterior offer superb views of the ocean. The roof, on which plants grow through Portuguese limestone cobblestones, makes up a large curved public square, and two glass monoliths echo the two rocky outcrops standing in the ocean that can been seen from Ilbarritz beach.

The City of the Ocean adopts an environmental

approach through its perfect integration into the landscape. Situated south of the town, below the château of Ilbarritz, the building is based on the principle of transparency. It is built around an open-air square with the sea for its horizon, a real call to the waves. Its environmental construction also incorporates a self-regulating heating mechanism, a system of rainwater collection via the cobblestone paving and a line of natural vegetation.

4

由纽约建筑师斯蒂芬·霍尔和他的巴西同事索兰格·法比奥所设计的海洋城效仿了巴斯克海岸的海浪造型。由这两位著名建筑师所打造的原始结构以"蓝天之下、碧海之下"为概念，采用了凹凸不同的几何体表面。

主楼内设有游戏科普区、展览区、礼堂和办公室。配有露台、自助餐区和点心吧的餐厅可以由外部直接进入，享有海洋的美景。屋顶上，植物在葡萄牙石灰鹅卵石的缝隙中生长。屋顶向下延伸，形成了巨大的弧形公共广场；两座玻璃结构与海上的两块礁石遥相呼应。

海洋城与景观环境完美地结合在一起。这座坐落在城市南部、伊巴利兹城堡脚下的建筑以通透感为设计原则。它围绕着一个露天广场建造，以大海为背景，宛如海浪。同时，它在环境建设上采用了自动调节的供暖系统、雨水收集系统（通过卵石铺地）和自然通风系统。

1-2. City of the Ocean at night
3. Aerial view
4. Exterior close-up view at night
5. Exterior staircase

1、2. 海洋城夜景
3. 鸟瞰景观
4. 夜晚外部特写
5. 外部楼梯

6. Detail of the roof terrace
7. East façade
8. Covered courtyard, west façade

6. 屋顶露台细部
7. 东立面
8. 西立面覆盖的庭院

9. Staircase leading to the roof terrace
10. Interior staircase
11. Entrance
12. Footbridge
13. Cafeteria
14. Bathyscaphe
15. Sub-marine base

9. 通向屋顶露台的楼梯
10. 内部楼梯
11. 入口
12. 行人桥
13. 咖啡馆
14. 深潜器
15. 潜艇基地

14

15

CARTIER FOUNDATION FOR
CONTEMPORARY ART

卡地亚当代艺术基金会

建筑师 Architect: Jean Nouvel
地点 Location: Paris
完工日期 Completion Date: 1994
摄影师 Photographer: Philippe Ruault

After 10 years in Jouy-en-Josas the Cartier Foundation moved to Boulevard Raspail, Paris in 1994. Here, Jean Nouvel created an ethereal structure of glass and metal whose transparency elongates the perspective of the boulevard. A magnificent Lebanon cedar planted in 1823 by Chateaubriand is echoed by a vertical garden above the entrance, created by the tropical botanical specialist Patrick Blanc in 1998. The garden and the architecture marry in this space without merging into one.

The ground floor is entirely "open" onto the garden, offering an extraordinary volume with its eight-metre ceiling height. Jean Nouvel sought a quasi-dematerialisation of the building, multiplying the possibilities in a flexible, free and dynamic space. Six floors are placed on top of this, of which three are

devoted to exhibitions. The changing appearance of the Foundation for each new exhibition thus calls out to passers-by.

"It's all about an architecture of lightness, windows and finely hatched steel. An architecture that blurs the tangible limits of the building and creates a poetic of mist and vanishing where the reading of a solid volume is redundant. When virtuality

attacks reality, architecture, more than ever, must have the courage to take on an image of contradiction," says Jean Nouvel, crystallising the essence of the building in a few lines.

1. Entrance
2. Exhibition space
3. Garden
4. Car park entrance
5. Offices

1. 入口
2. 展览空间
3. 花园
4. 停车场入口
5. 办公空间

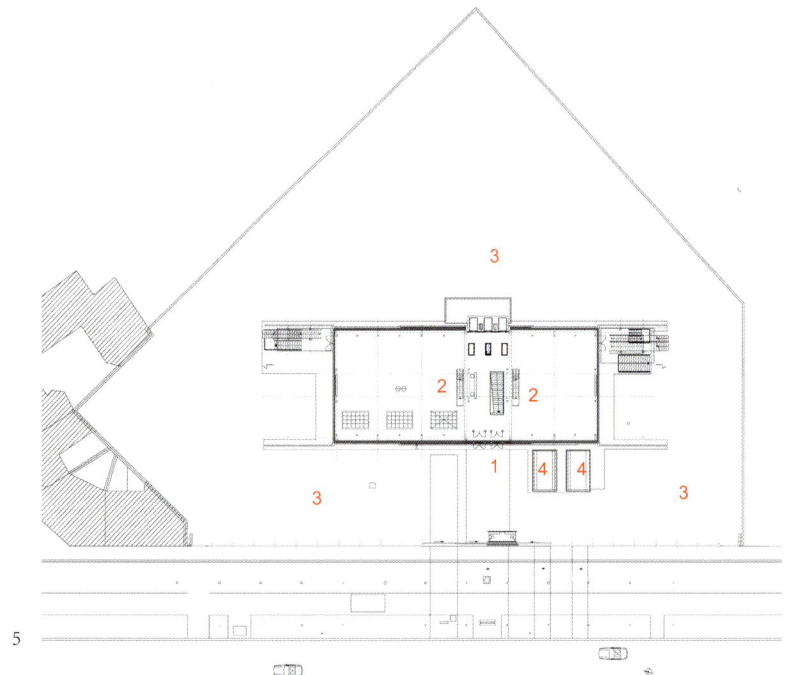

6

在茹伊昂若萨栖身十年之后，卡地亚基金会决定在1994年迁至巴黎的拉斯帕丽大道。让·努维尔打造了一座缥缈的玻璃金属结构，它的通透感拉长了街道的远景。一棵由夏多布里昂种植于1823年的黎巴嫩雪松在入口处与帕特里克·布兰克于1998年设计的热带植物园相互辉映。花园和建筑相互结合，又各自独立。

建筑一楼彻底地面向花园敞开，8米高的高度打造了非凡的空间体验。让·努维尔在建筑中追求一种类似去物质作用的方式，放大了灵活、自由、动感空间的可能性。上面的六层楼中有三层是展览空间。这样一来，基金会的每次新展览都能吸引着过往行人。

在描述建筑的精华时，让·努维尔说道："这座轻盈的建筑结合了诸多窗口和精制的钢材。建筑模糊了有形的界限，打造了富有诗意的模糊感，让实体结构变得多余。当虚拟碰撞现实，建筑必须拿出足够的勇气来呈现出矛盾的形象。"

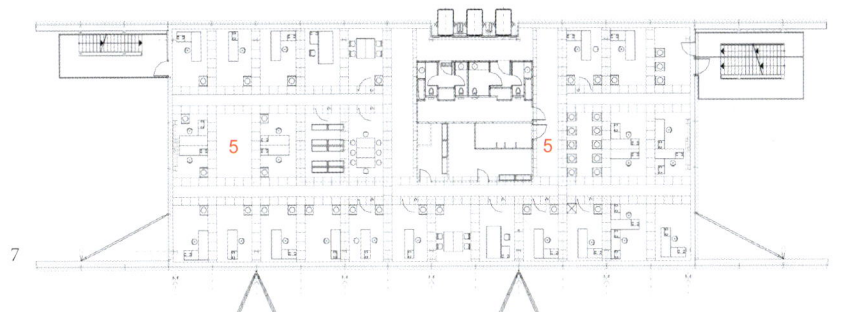

7

8. View from Boulevard Raspail
9. César exhibition
10. Cross section

8. 从拉斯帕丽大道看向展览厅
9. 艺术家凯撒作品展
10. 横向剖面图

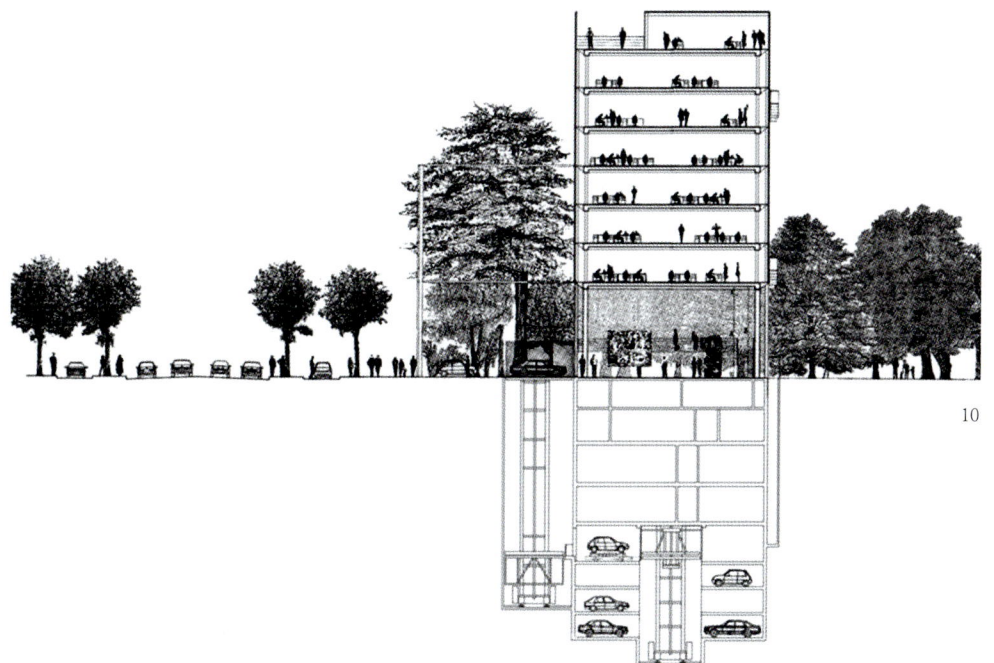

MEAUX COUNTRY MUSEUM OF THE GREAT WAR

莫城一战博物馆

建筑师 Architect: Christophe Lab
地点 Location: Meaux
完工日期 Completion Date: 2011
摄影师 Photographer: Philippe Ruault

The building of the Meaux Country Museum of the Great War is a brutal form, symbolising the effect of the mobilisation of forces on humankind and on the landscape. This allusive form disrupts the geography, concentrating the energy of war. The reference to the tortuous field of battle, the uprooted ground, is not literal but impregnates the imagination.

A sheltered forecourt leads to the entrance of the museum. It is a covered space, a transition between the exterior and the interior. Its light grey paving is a vast silent map, representing the north-east quarter of France and reflecting a soft light. It can receive large projections showing the movements of troops at different points in the war. Crossing the forecourt to the entrance hall, the visitor's path is illuminated and punctuated by pools of light shed by luminous openings like

enormous "spotlights" overhead. Looking up they are intrigued by elements of the museum displays on the first floor partially visible through these glass openings.

A staircase with deep treads and shallow risers takes the visitor to the reception's upper level. It climbs around the shop, which the visitor perceives from different angles before reaching the ticket desk and then the exhibition itself. Neutral

mannequins appear to walk out of display cases 14 and 18, dissolving the glass screens in order to limit their significance and offering a spectacular run up to a "heightened" reality. Here, sound, moving light projections and partial mirrors doubling the crowd effect evoke the experience of war and throw the spectator in among the mannequins.

1. Façade on Route de Varredes
2. Trench suspended from beneath the building
3. Main façades and forecourt
4. Master plan
5. Approach from the forecourt

1. 面向瓦雷德路的建筑立面
2. 建筑下方悬浮的沟槽
3. 主立面和前庭
4. 总规划图
5. 前庭小径

莫城一战博物馆的建筑造型十分粗犷，象征了武力对人类和景观的影响。这个富有暗喻性质的造型扰乱了地势，浓缩了战争的能量。参照了扭曲的战场，建筑场地夸张地渗透了这一形象。

受遮蔽的前庭直通博物馆的入口。这个带顶的空间是室内外的过渡。它的浅灰色路面是一幅巨大而无声的地图，呈现出法国东北部地区，反射出柔和的光线。它可以接收大型投影，反映战争时期军队的移动。从前庭进入门厅，参观者的路线都被头顶的巨型"聚光灯"所指引。向上看，他们会被博物馆二楼玻璃窗所展现的丰富藏品所吸引。

宽阶楼梯带领着参观者进入上层的前台接待区。楼梯围绕着商店缓缓上升，让参观者可以在买票进入展览之前各个角度观察商店。人体模型似乎从14和18号展示柜中缓缓走出，隐去了玻璃屏风，带领人们走进"升华的"现实。声效、引动的光线投影和半透镜让人群效果加倍，突出了战争体验，将参观者抛进了人体模型之中。

6. Access to the upper reception area
7. Crossroads of the central hub: the arrival and exit staircases
8. Arrival at the upper reception area, shop and ticket office

6. 通往上层接待区
7. 中心枢纽交叉口，入口和出口楼梯
8. 抵达上层接待区、商店和售票处

9

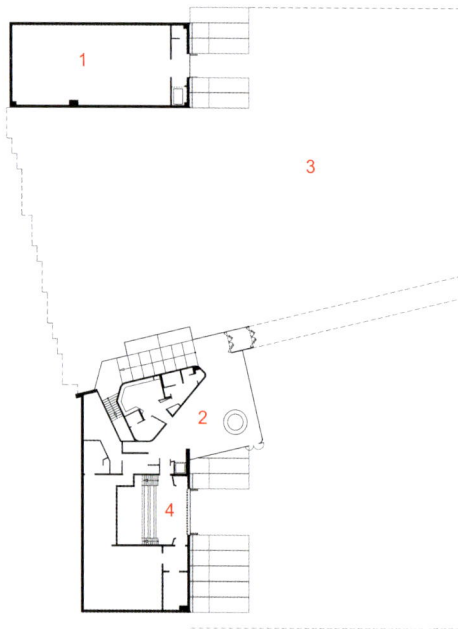

10

11

1. Lower reserves
2. Reception and snack bar
3. Forecourt
4. Auditorium
5. Upper reserves
6. Administration
7. Reception and shop
8. Educational workshops
9. Temporary exhibitions
10. Central nave
11. Thematic exhibitions

1. 底层储藏室
2. 接待处与餐饮吧台
3. 前庭广场
4. 演映庭
5. 高层储藏室
6. 行政空间
7. 接待处与商店
8. 教学工作室
9. 临时性展览区
10. 中央大厅
11. 主题性展览区

12

9. Upper reception area
10. Lower ground floor plan
11. Upper ground floor plan
12. Shop and ticket office
13. Cross section AA
14. Cross section BB

9. 上层接待区
10. 下层平面图
11. 上层平面图
12. 商店和售票处
13. AA横截面
14. BB横截面

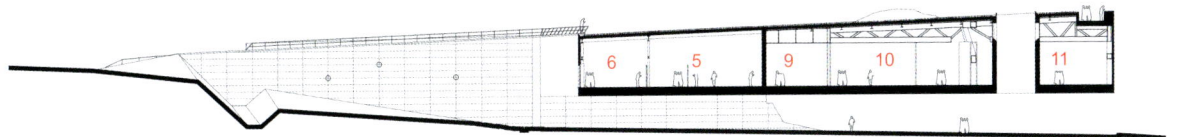

13

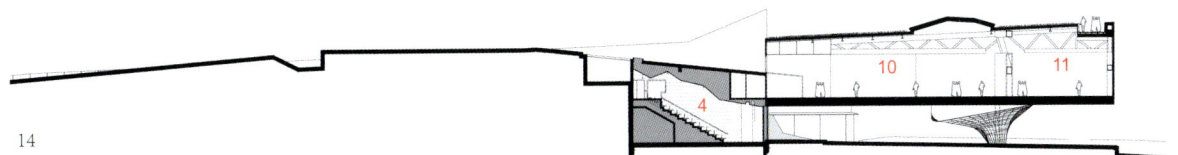

14

15. Introductory room with a fresco created by Jacques Tardi
16. "From school to the dugout" area
17. Marne 18 display case, from which a "neutral" mannequin emerges
18. German trench

15. 介绍室的壁画由雅克塔蒂所创作
16. "从学校到防空洞"展览区域
17. 马恩河18号展示柜，从中走出一位"中性"的模特
18. 德国战沟

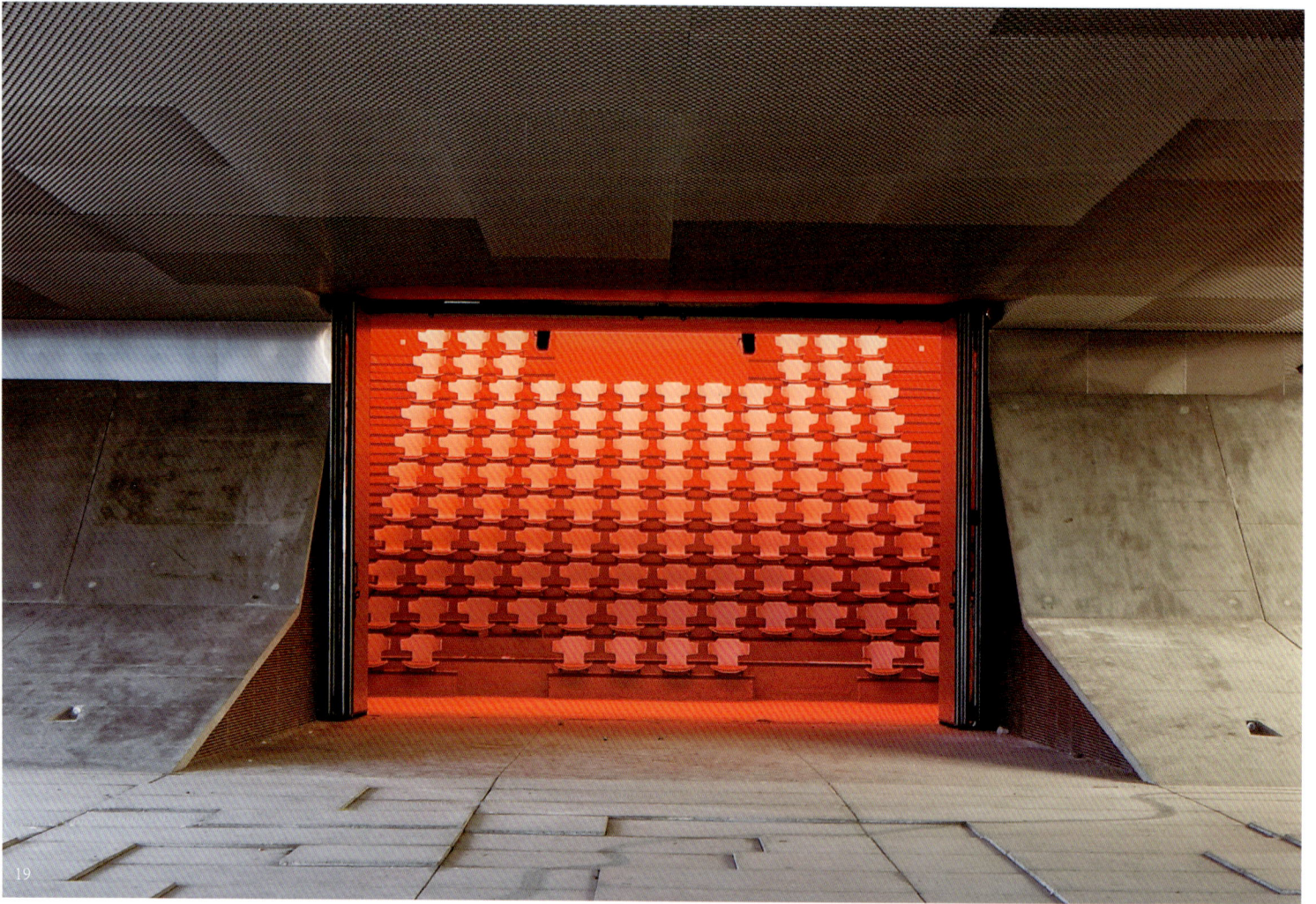

19-20. Auditorium seen from the
forecourt
21. Forecourt seen from the auditorium
22. Auditorium section, backstage open
to the forecourt

19、20. 从前庭看到的礼堂
21. 从礼堂观众席看到前庭
22. 礼堂剖面图，后排直通前庭

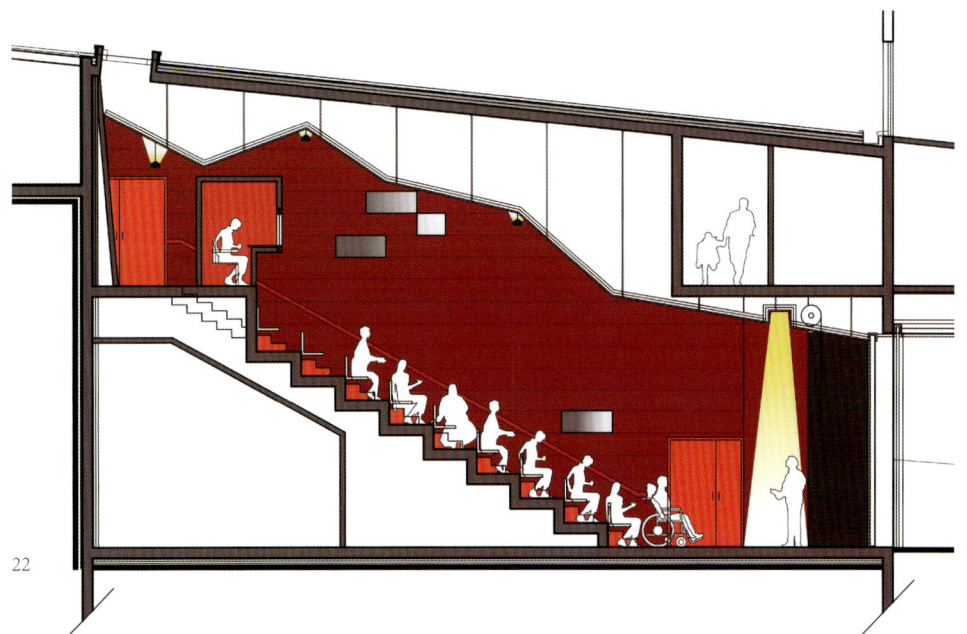

INSTITUTE OF THE ARAB WORLD
阿拉伯世界文化中心

建筑师 Architect: Jean Nouvel, Pierre Soria, Gilbert Lezenes, Architecture Studio
地点 Location: Paris
完工日期 Completion Date: 1987
摄影师 Photographer: Philippe Ruault

The Institute of the Arab World was opened in 1987, the fruit of an agreement between France and 19 Arab countries, later joined by three more, to encourage a greater understanding of the contemporary Arab world and Arab-Muslim culture. On the banks of the Seine, facing the Ile Saint-Louis and adjoining Jussieu University, it combines an imaginative synthesis of architectural ideas from the Arab world and the West.

The spirit of Arab culture is particularly evoked in the diaphragms placed on the south façade, 240 moucharabiehs that open and close according to the outside light, thus creating a play of light and shadows inside the institute. Light has great importance here, as Jean Nouvel explains: "I started to consider the question of light at the Institute of the Arab World. The theme of light is seen on the south wall, composed only of photographic diaphragms, in the

organisation of the staircases, the flow of contours, superimpositions, reflections and shadows." The north façade is "a literal mirror of western culture", reflecting, but also reproducing, the landscape of Paris through a series of lines, of signs capturing the spirit of Paris, an echo of the buildings on the opposite banks.

The entrance to the auditorium, below ground, is pure fantasy: the space seems infinite, while the repetition of massive,

strong columns recalls the hypostyle rooms of Arab buildings. The tower of books, the patio and the interior courtyard also show the positive results of Jean Nouvel's questioning of western culture, which, having distanced itself from all clichés, finds in this building a strong and shared identity.

1. View from the Seine
2. South façade
3. North façade, detail
4. West façade, interior of the tower of books
5. Patio, detail
6&8. South façade, interior
7. Axonometry of the museum side
9. Ground floor plan

1. 从塞纳河看阿拉伯世界文化中心
2. 南立面
3. 北立面细部
4. 西立面，藏书塔内部
5. 中庭细部
6、8. 南立面室内
7. 等角透视
9. 一楼平面图

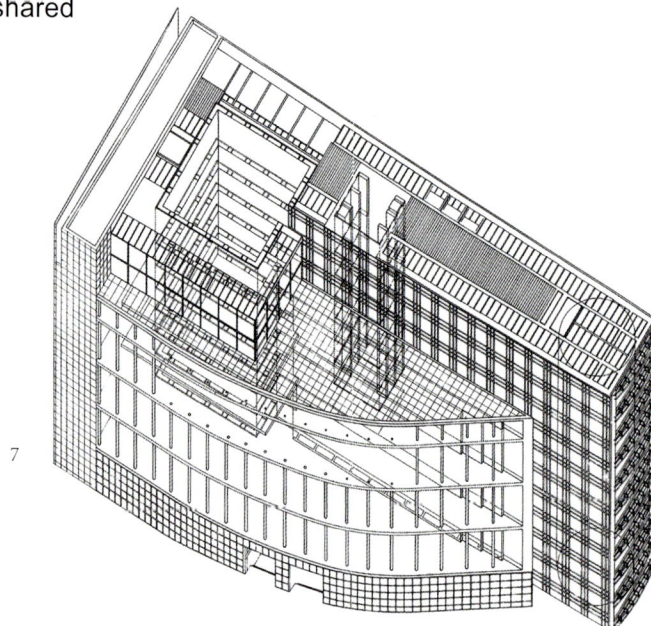

1. Forecourt
2. Entrance
3. Entrance hall

1. 前庭广场
2. 入口
3. 接待大厅

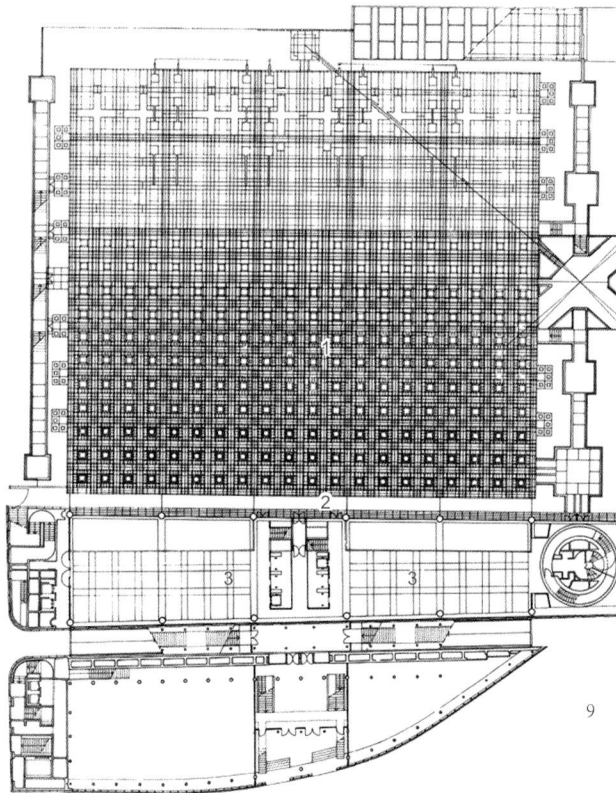

阿拉伯世界文化中心开放于1987年，是法国与19个阿拉伯国家（后来又加入了3个）一致同意的成果，旨在促进人们对当代阿拉伯世界和穆斯林文化的理解。文化中心位于塞纳河畔，面朝圣路易岛，并且紧邻巴黎第六大学，综合了阿拉伯世界和西方世界的建筑形态。

阿拉伯文化的精髓被特别地体现在建筑南立面的隔板上：240个木雕窗格根据外部光线而闭合，在文化中心内打造出奇妙的光影效果。正如建筑师让·努维尔所说，光线在设计中至关重要："我开始思考光线在阿拉伯世界文化中心中的作用。光线的主题呈现在南墙上，由逼真的隔板、楼梯结构、轮廓、叠加、倒映和影子组成。"北立面是"西方社会的一面镜子"，通过一系列线条、捕捉巴黎精神的标志以及与对岸建筑的呼应反映并再现了巴黎的景观。

设在地下的礼堂入口展现出奇妙的效果：空间看起来无穷无尽，巨大的立柱的重复让人们想起了阿拉伯建筑中多柱的房间。书塔、天井和室内庭院都是让·努维尔对西方文化探寻的正面结果，让建筑远离了老生常谈，树立了强烈的共享形象。

10

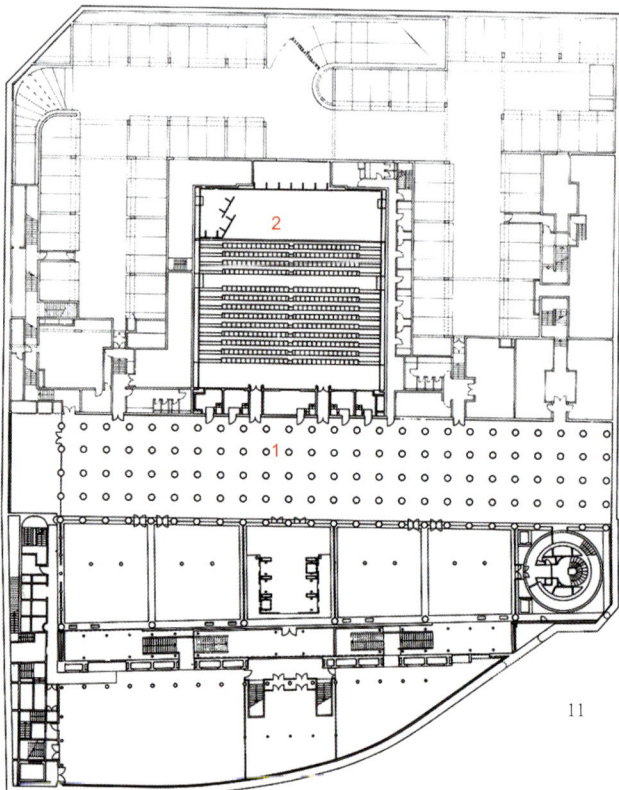

11

10. Smoking room
11. Basement floor plan
12. Staircase / south façade
13. 5th floor plan
14. 6th floor plan
15. 7th floor plan
16. 9th floor plan

10. 吸烟室
11. 地下室平面图
12. 楼梯间和南立面
13. 六楼平面图
14. 七楼平面图
15. 八楼平面图
16. 十楼平面图

1. Hypostyle room   1. 多柱式大厅
2. Auditorium   2. 表演厅 / 讲堂
3. Museum entrance   3. 博物馆入口
4. Museum   4. 博物馆展厅
5. Museum atrium   5. 博物馆展厅挑空
6. Offices   6. 办公室
7. Library   7. 书店
8. Library atrium   8. 书店挑空
9. High council room   9. 高层会议厅
10. Restaurant   10. 餐厅
11. Terrace   11. 露天平台

12

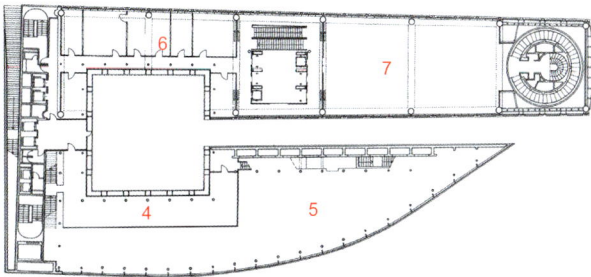

13

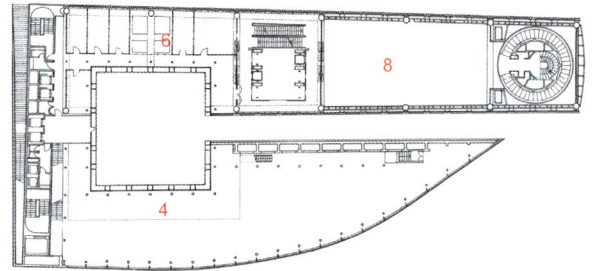

14

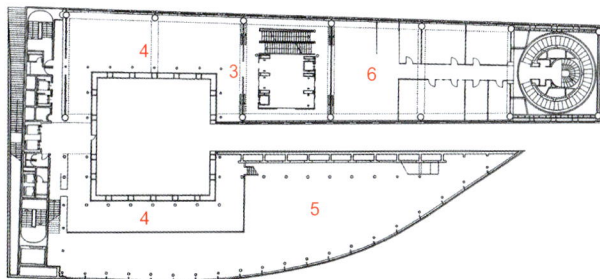

15

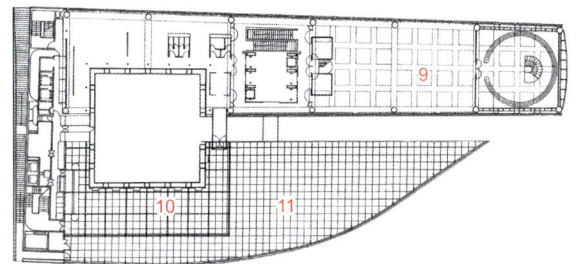

16

MOBILE ART PAVILION
移动艺术馆

建筑师 Architect: Zaha Hadid
地点 Location: Paris
完工日期 Completion Date: 2011
摄影师 Photographer: Francois Lacour,
courtesy of Institut du Monde Arabe
线图 Drawings: Zaha Hadid Architects

After being shown in Hong Kong, Tokyo and New York, the Mobile Art Pavilion designed by Zaha Hadid was donated to the Institute of the Arab World in Paris by Chanel, where it will form a permanent exhibition venue. Its first exhibition, running from 29 April to 30 October 2011, was on the work of Zaha Hadid herself, offering a three-fold entry into her world: through the building, the exhibition design and the projects exhibited. The aim was, she says, to evoke "a kind of strangeness and newness that is comparable to the experience of going to a new country".

The pavilion illustrates Hadid's use of digital imaging technology to break free of the geometric constraints of the industrial age. Its organic form has evolved from the spiraling shapes found in nature. This system of organisation and growth, expanding towards its circumference, gives the pavilion generous public areas at its entrance with a 125m^2 terrace. Through the parametric distortion of a torus, a constant variety of exhibition spaces are

created around its circumference, whilst at its centre, a large 60m² courtyard with natural lighting provides an area for visitors to meet and reflect on the exhibition. This arrangement also allows visitors to see each other moving through the space, facilitating the viewing of art as a collective experience.

The organic fibre reinforced plastic shell is created with a succession of reducing arched steel segments. As the pavilion has traveled over three continents, this

segmentation allows the structure to be easily transported. The partitioning seams become a strong formal feature of the exterior façade, whilst creating a spatial rhythm of perspective views within the interior exhibition spaces.

In creating the Mobile Art Pavilion, Zaha Hadid has developed the fluid geometries of natural systems into a continuum of fluent and dynamic space – where oppositions between exterior and interior, light and dark, natural and artificial landscapes are synthesised. Lines of energy converge within the Pavilion, constantly redefining the quality of each exhibition space whilst guiding movement through the exhibition.

1. Night-view Perspective of Mobile Art Pavilion in front of the Institute of the Arab Word (Institut du Monde Arabe – IMA)
2. Birdseye Perspective of the Mobile Art Pavilion by night illuminated via RGB machine
3. Birdseye Perspective of the Mobile Art Pavilion by day
4. Roof plan
5. Ground floor plan

1. 位于阿拉伯世界文化中心前庭广场的移动艺术馆夜景
2. 由三色机照亮的移动艺术馆鸟瞰夜景
3. 移动艺术馆的日间鸟瞰景观
4. 屋顶平面图
5. 一层平面图

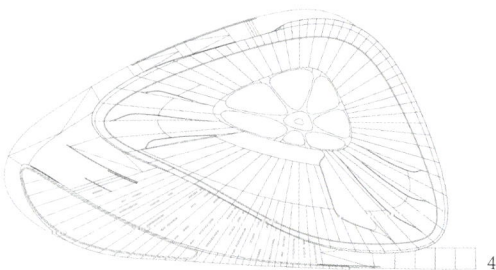

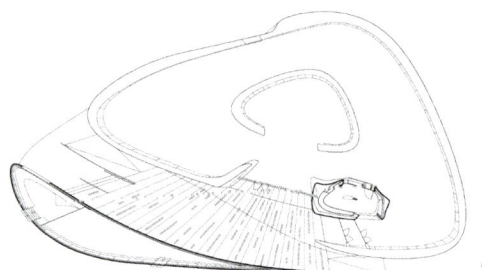

6. Eyelevel perspective of Mobile Art pavilion in front of the IMA
7. Front Elevation
8. Back Elevation
9. Right Elevation
10. Left Elevation

6. 移动艺术馆与阿拉伯世界文化中心
7. 前立面
8. 后立面
9. 右立面
10. 左立面

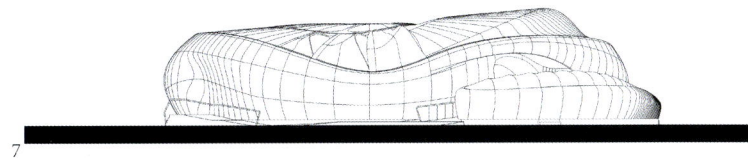

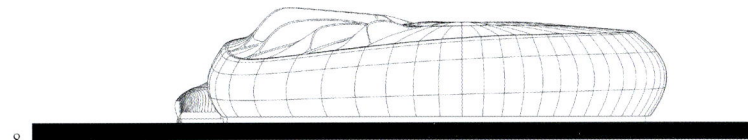

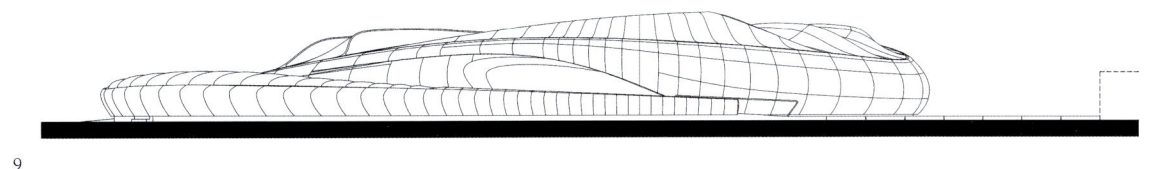

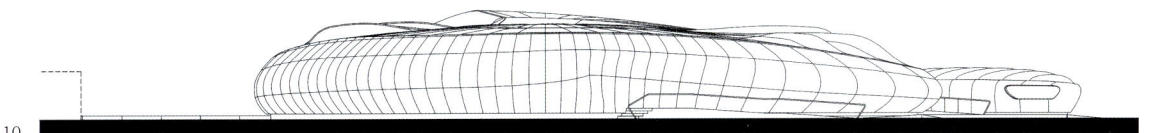

11

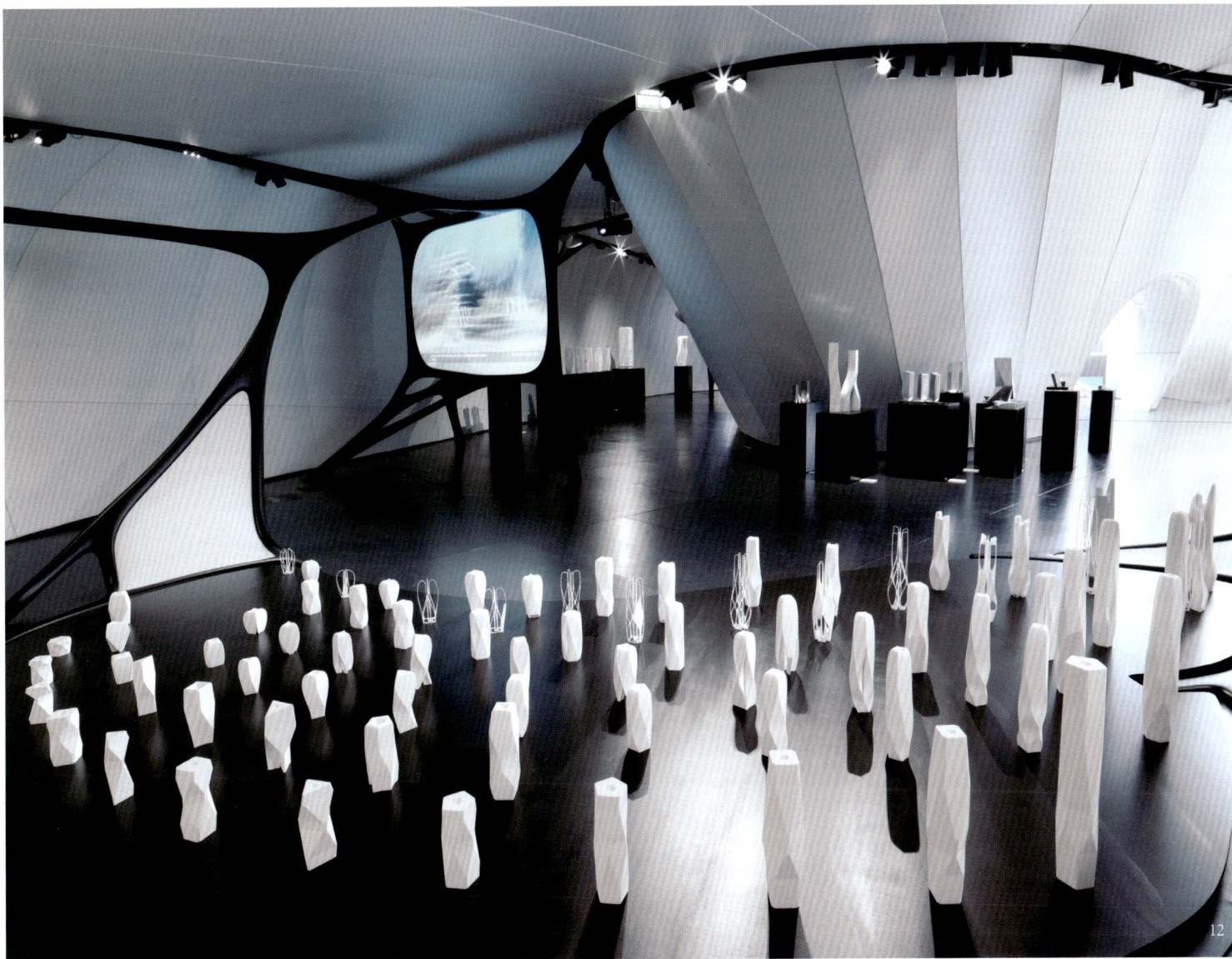

11. Interior View of exhibition Installation "Zaha Hadid – une architecture" with 3-dimensional scenographic structure housing large projection screens
12. Interior view towards the courtyard with Towerfield of Beijing Central Business District Competition in the foreground and silver models in the background

11. 扎哈·哈迪德设计展览内景——采用三维立体结构和大型投影屏幕
12. 前方是北京中央商务区竞赛项目的各种塔形，后方是银色模型

在香港、东京和纽约巡回展出之后，由扎哈·哈迪德设计的移动艺术馆被捐赠给了巴黎阿拉伯世界文化中心，在其前庭广场落脚，成为一个永久性展览场所。它的首次展览从2011年4月29日延续到10月30日，展出了扎哈·哈迪德自己的作品，从三个方面呈现了她的世界：建筑、展场设计和项目展示。据哈迪德说，展览的目的是为了唤起"一种仿佛前往新国度的奇异感和新鲜感"。

展馆的设计展示出哈迪德是如何运用数位图像技术来打破工业时代几何形体的限制。它的有机造型从自然界的螺旋型体演化而来。这一系统沿着自身的周长延伸，为展馆在入口处（设有125平方米的平台）提供了宽敞的公共空间。通过环面的扭曲，各种各样的展览空间沿着圆周展开。展馆中心一个60平方米的自然采光庭院为参观

者提供了会面和反思展览的空间。这种布局还让参观者可以在穿越空间时看到彼此，使得艺术品的观赏成为一种集体共创的经验。

有机纤维强化塑料外壳由一系列的拱形钢元件支撑。在展馆游历三个大陆的旅程中，这些元件保证了结构的便利运输。外壳的分割接缝成为立面上的强烈特征，也在内部展览空间里为人们的视线创造出一种特殊的空间旋律。

在移动艺术馆的设计中，扎哈·哈迪德将大自然系统里的流线几何融入了连续的动态空间，并将室内与室外、光明与黑暗、自然景观与人造景观等对比效果相互结合。充满活力的流线在展馆中汇聚，不断地重新定义每个展览空间的特色，也引领着人们在展馆中的移动。

13

14

13. Interior view with Lunar Relief mounted in front of textile skin and the Slovakian Bratislava City Center Model and Animation as well as the Moroccan Rabat Tower exhibited on black plinths
14. Detail Lunar Relief
15. Interior view of 3-dimensional cellular structure made from CNC milled polyurethane foam painted in matt black with a car paint system and stretch fabric suspended from cells as projection devices
16. The pavilion roof topography was modeled on a spherical geometry using a radial grid
17. The inner walls of the Pavilion are made from a stretch fabric that was developed through a series of cutting patterns

13. 展馆内景："月球表面"正对着纺织墙面和斯洛伐克的布拉迪斯拉发城市中心大厦的模型和动画，以及位于黑色底座上的摩洛哥的拉巴特塔模型
14. "月球表面"细部
15. 展馆内景：用汽车涂漆涂成黑色亚光的聚氨酯泡沫材质，以数控系统组成三维蜂窝结构体
16. 展馆屋顶以辐射状网格的球面几何作为模板
17. 展馆内墙使用的弹力织物通过一系列的切割模式开发而成

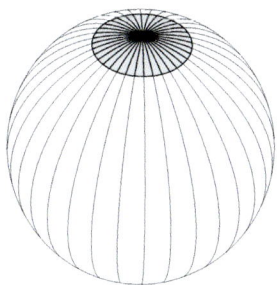
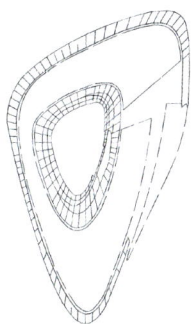
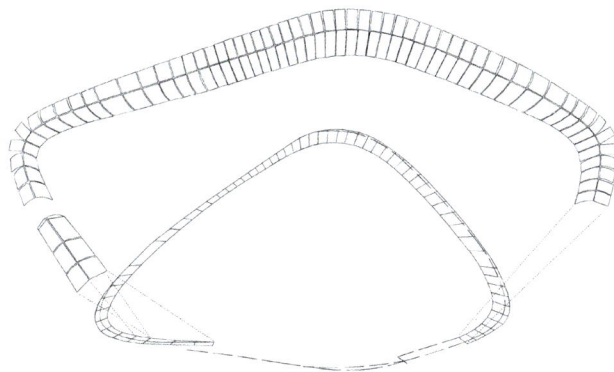

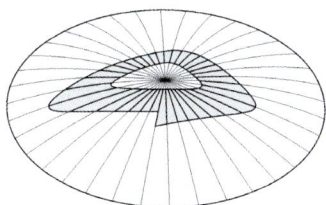
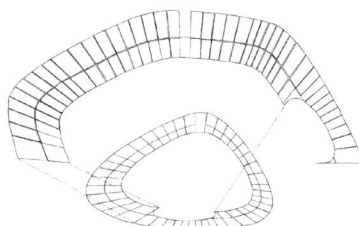
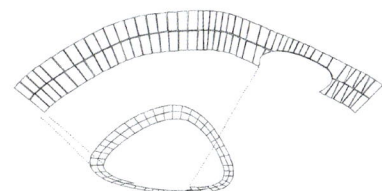

15

16

17

18

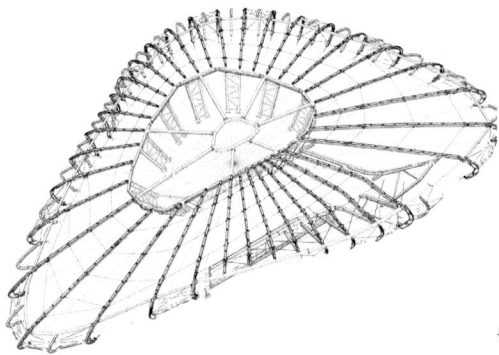

19

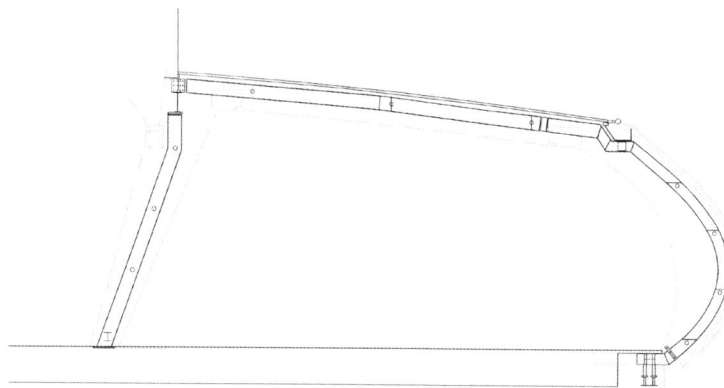

20

18. Detail of scenographic elements enveloping 6 large scale Tower models CNC milled from foam
19. Radial structural arches composing the primary steel structure of the pavilion
20. Dimensioned radial section of an arch composed of a series of bend I-Beams
21. Detail of the Abu Dhabi performing Arts Center embedded within the fluid environment of the pavilion and its scenographic design
22. Detail of Silver Painting -Barcelona Tower- floating in front of textile internal skin
23. Detail of Silver Models and projection screen

18. 展览场景的构成元素包围着以数控系统和泡沫材质制成的6座大型塔楼模型
19. 放射状的拱型结构组成展馆的主要钢架
20. 由一系列弯曲的I型梁组成的规格统一的放射状拱型结构
21. 展馆的流线结构和场景设计美妙地衬托出阿布扎比表演艺术中心的模型
22. 巴塞罗那大楼的银色模型绘画，浮挂在展馆内部的纺织墙上
23. 银色模型和投影屏幕

JEAN COCTEAU MUSEUM

让·科克托博物馆

建筑师 Architect: Rudy Ricciotti
地点 Location: Menton
完工日期 Completion Date: 2011
摄影师 Photographer: Olivier Amsellem

Winner of an international competition launched by the town of Menton in 2008, Rudy Ricciotti's project will house the Séverin Wunderman donation in its entirety in a 2,700m² space. The site is not neutral and the project entails strengthening the town out towards the sea, creating a bedrock to ensure that it will sit on the existing "urban canvas". The building forms part of the urban framework but the quay on which it is situated had to be reclaimed for pedestrians. In the distance the cloudy vision of the arcades,

an image from old postcards, reminds us of Menton when it was a fashionable seaside resort.

The museum does not try to hide the sea, but is rather a plinth and architectural narrative, forging a new link with the original, elegant era of Menton where the great stylistic movements of the 1900s were played out. The building must convey the freedom of its romanticism: it is there because it could not be elsewhere. It has its place just as the covered market has

its place. It is an urban form that helps to repair a fragment of this neighbourhood, to convey the fact that this now urban space was once just an embankment on the sea. The building must let itself be discovered. Should it not retain a mystery? The mystery of its constructive truth, of its stasis? The museum accepts its appearance, intrigues through its transparency, attracts by what it lets us glimpse. The architecture summons up the elusive and complex world of Jean Cocteau.

1-2. The black and white aesthetic is linked to the works of Jean Cocteau
3. The museum has sweeping views of the exterior
4. The arcades, evoking memories from old postcards
5. Master plan
6. Transition space between interior and exterior

1、2. 博物馆的空间的黑白美学与让·科克托的作品紧紧相关
3. 一个向外敞开的博物馆
4. 柱廊建筑物——从旧明信片中唤起的回忆
5. 总规划图
6. 内部和外部的过渡

7

8

9

10

7-9. Interior views: the play of shadows creates sublime effects
10. A museum as a mirror of the artist and his work

7-9. 内部景观：优雅细致的光影效果
10. 博物馆就像是艺术家和其作品的反射镜

Dreaming up this museum was like imagining an architectural principle that could take the contrast between light and darkness and make it into a sublime play of shadows. The architectural approach, and above all the aesthetic of black and white, convey the dream-world, the mystery and the complexity of the works of Jean Cocteau. A museum as a mirror of the artist and his work. Which means refusing the dictature of a tyrannical modernity and considering story, dreaming and design as a possible opening for architecture.

作为2008年芒通市举办的国际设计竞赛的优胜者，鲁迪·里西奥第的项目将在2,700平方米的空间内安置赛弗林·温德曼所捐赠的藏品。项目必须强化城市与海洋之间的关系，打造一个基岩以保证它能够栖身于原有的城市布局之中。建筑形成了城市框架的一部分，但是其所在的码头必须被改造成步行专用。从远处看，旧明信片上常见的拱形柱廊让我们想起了芒通还是时髦的海滨度假村的时代。

博物馆并没有试图隐藏海洋，而是作为一个基石，以建筑的形式将今日的芒通与20世纪初曾走在时尚运动尖端的芒通联系起来。建筑必须慷慨洋溢着浪漫精神，因为它只属于这里。它犹如一个带有遮顶的传统市场，在城市里必然有属于它的位置。它的城市造型帮助修复了周边地区的空缺，传达出：现在的城市空间曾仅仅是海岸而已。建筑必须展现自我。它该不该保持神秘？它建造的秘密，静止的秘密。博物馆接受了自己的外观并以通透感激发人们的兴趣。建筑唤起了让·科克托那晦涩难懂的复杂世界。

构思这座博物馆的过程就像是构造一座明暗对比建筑，使其在光影中升华。建筑的构架方式以及黑白两色的运用表现出梦幻世界、神秘感和让·科克托作品的复杂性。博物馆就像是艺术家和他作品的镜子，意味着它拒绝了现代化的统治，将故事、梦想和设计一起融入了建筑之中。

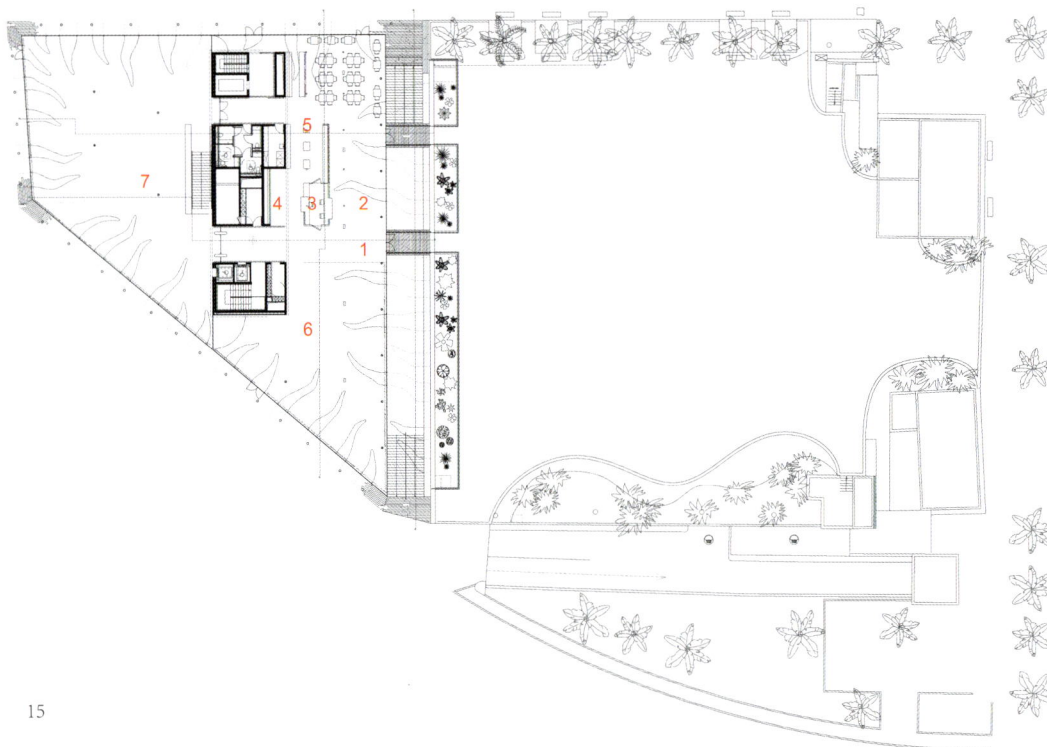

1. Entrance
2. Hall
3. Reception
4. Bookshop
5. Café
6. Temporary exhibition
7. Permanent exhibition
8. Reserve
9. Museographical workshop
10. Maintenance and Upkeep Workshop
11. Graphic arts space
12. Documentation
13. Offices
14. Educational workshop
15. Car park

1. 入口
2. 大堂
3. 前台
4. 书店
5. 咖啡厅
6. 临时展区
7. 常设展区
8. 储藏区
9. 博物馆工作室
10. 维修和保养车间
11. 平面艺术空间
12. 档案室
13. 办公室
14. 教育工作室
15. 停车场

11-14. The importance of contrast
between light and shade
15. Ground floor plan
16. The architectural device of the
arcades has been revisited in an entirely
contemporary spirit
17. 1st floor plan

11–14. 大量的光影对比效果
15. 一层平面图
16. 老式柱廊建筑被赋予了全新的现代感
17. 二层平面图

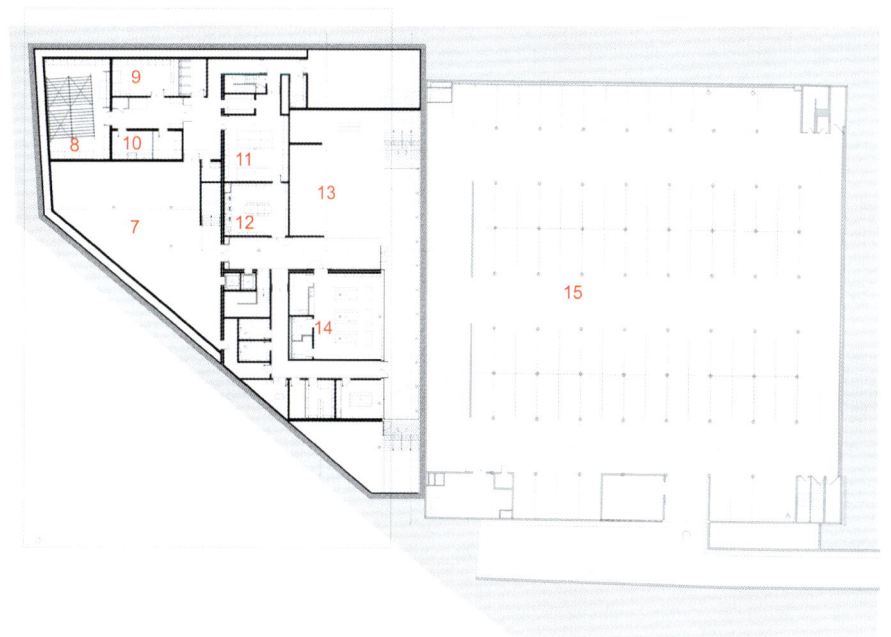

VAL-DE-MARNE MUSEUM OF CONTEMPORARY ART

瓦勒德马恩当代艺术博物馆

建筑师 Architect: Jacques Ripault
地点 Location: Vitry-sur-Seine
完工日期 Completion Date: 2005
摄影师 Photographer: Pauline Turmel (pp.248-249), Patrick Muller (pp.250-251), Jean-Marie Monthiers (pp.253-257)

The Val-de-Marne Departmental Museum of Contemporary Art, known as the Mac Val, gathers together the collection of paintings, sculptures and drawings by contemporary artists that have been acquired by the department of Val-de-Marne since 1982. The museum aims to make art and culture available to everyone, to create a place for learning, familiarisation and knowledge, but also a place for feeling and finding peace.

With an architecture that revolves around the works, the artists and the public, this museum offers an opportunity for self-development and fulfilment.

The Mac Val is entered at street level via a longitudinal gallery, with a pathway that leads to the garden. Moving from east to west, the visitor encounters the amphitheatre, the restaurant and its terrace. In the centre, near to the entrance

1

hall, is found a bookshop and on the upper floor a library. In the west wing a wide, light-filled corridor opens onto large rooms lit via skylights in grooves running the length of the ceiling. In the south wing, a hall free of structures uses light reflectors. The museum was designed not to appear monumental, imposing or spectacular at first sight. It offers a linear geometry, and contrasting planes and transparencies that dilate or draw in the spaces.

2

3

Light is the raw material, it carves up the volumes, slips between two planes, reflects itself between two partitions and marks contrasts over the course of the visit. The aesthetic is not one of stacking but of stretching out, with a construction of horizontal and vertical planes alternating with the light, which display paintings and sculptures and strengthen the lines of the public space. Through the contrasts of sand-coloured concrete and the dark wenge wood floors, this museum is essentially black and white; it is the works themselves that bring free forms and colours.

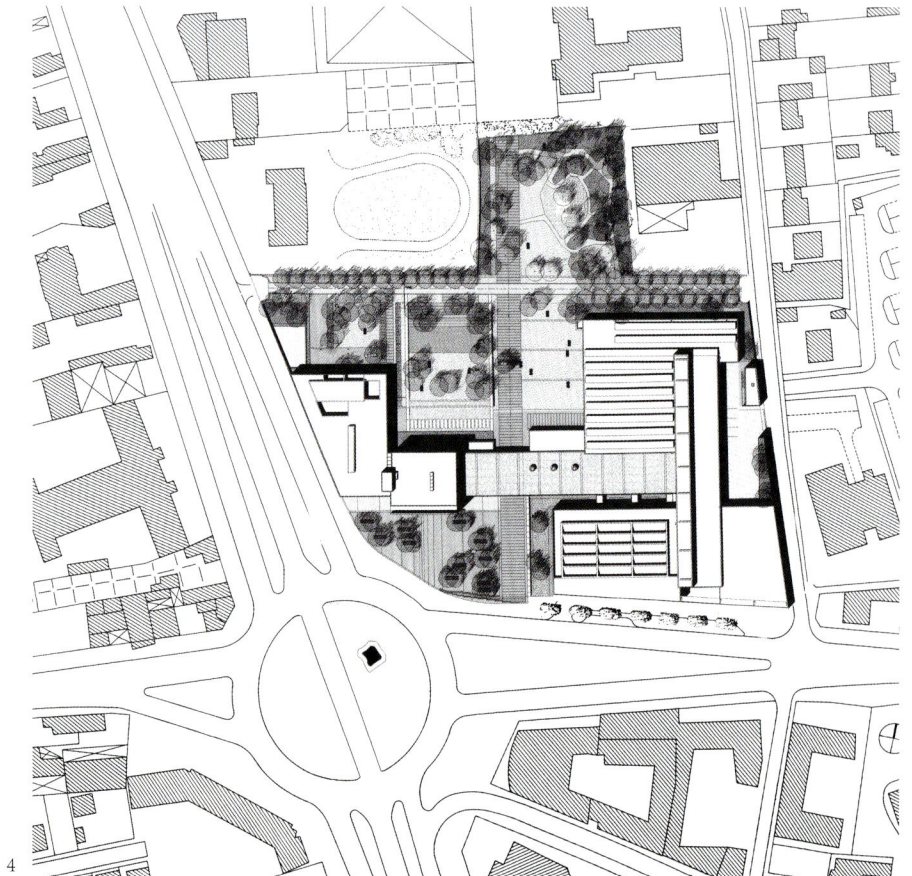

4

5

1. Overall view in its context
2. Grand gallery enfilade
3. Horizontal planes of the entrance
4. Master plan
5. Restaurant and artists' studios
6. Section

1. 博物馆全景
2. 连接多种空间的长廊
3. 入口处的水平线条
4. 总规划图
5. 餐厅和艺术家工作室
6. 剖面图

瓦勒德马恩当代艺术博物馆收藏了自1982年以来瓦勒德马恩所收集的当代艺术家的油画、雕塑和绘画作品。博物馆的目标是让人人都能接触到艺术与文化，打造一个学习、熟识、了解并能够令人感到平静的场所。博物馆的建筑设计围绕着艺术品、艺术家和公众展开，提供了自我发展、自我实现的机会。

参观者从临街的长廊进入博物馆，旁边的小路通往花园。自东向西，是圆形剧场、餐厅和餐厅露台。博物馆中央，紧邻门厅处是一家书店，楼上则是图书馆。博物馆西侧一条宽敞明亮的走廊通往带有天窗的大房间。南侧的大厅采用了反光灯。博物馆的设计并不宏伟壮观，它的直线结构、对比鲜明的平面和通透性让空间自由地扩大缩小。

光是一种原材料，它能够分割空间、在两个平面之间滑动、在两个隔断之间自我反射并且在参观路线中形成对比。博物馆的美不是堆叠而成，而是通过垂直和水平平面与光的交替伸展开来，让博物馆能够更好地展示绘画和雕塑作品，并且突出公共空间的线条。沙色混凝土和黑鸡翅木地板的对比让博物馆以黑白两色为主色调；而展览的艺术品则为博物馆带来了自由造型和色彩。

6

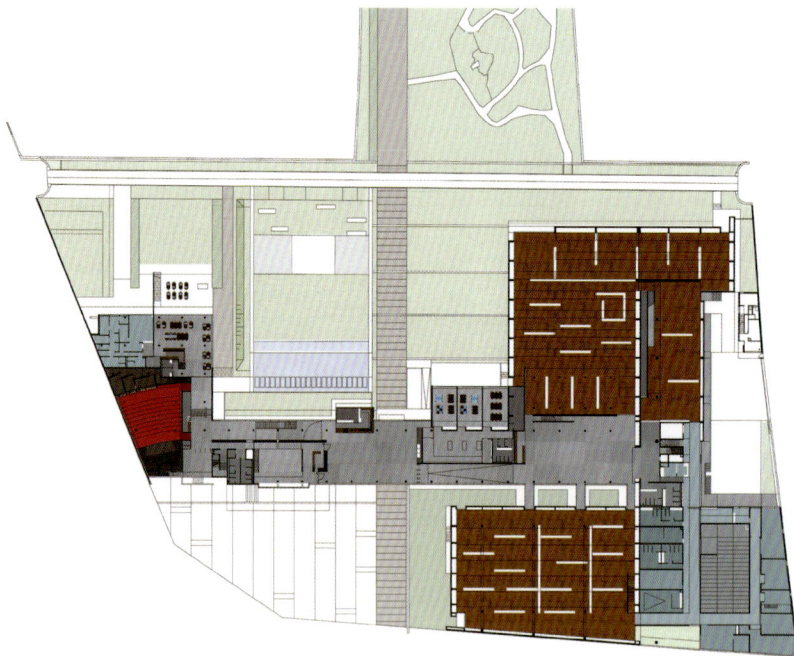

7

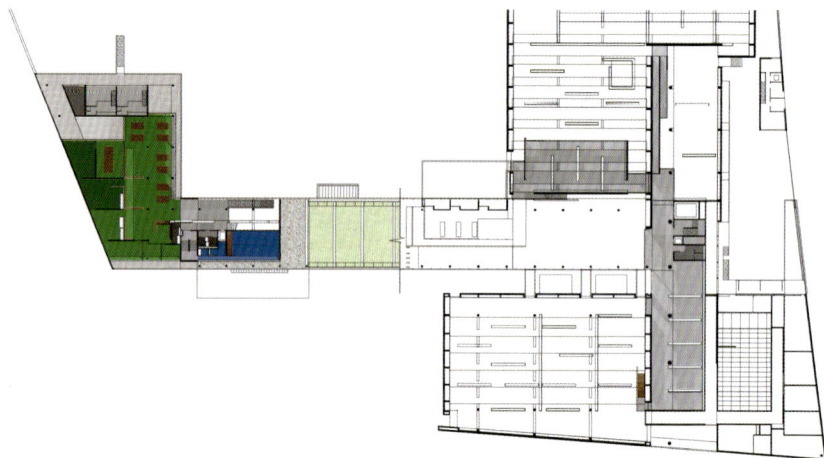

8

1. Educational spaces
2. Permanent exhibitions
3. Temporary exhibitions
4. Reserve   Restaurant
5. Auditorium – Research centre
6. Bookshops – Administration

A. Museum forecourt
B. Sculpture garden
C. Restaurant terrace
D. Delivery area

1. 教学空间
2. 永久性展览空间
3. 临时性展览空间
4. 储藏空间 – 修复空间
5. 大讲堂 – 资料中心
6. 书店 – 行政空间

A. 临时博物馆前广场
B. 雕塑花园
C. 餐厅露台
D. 送货区

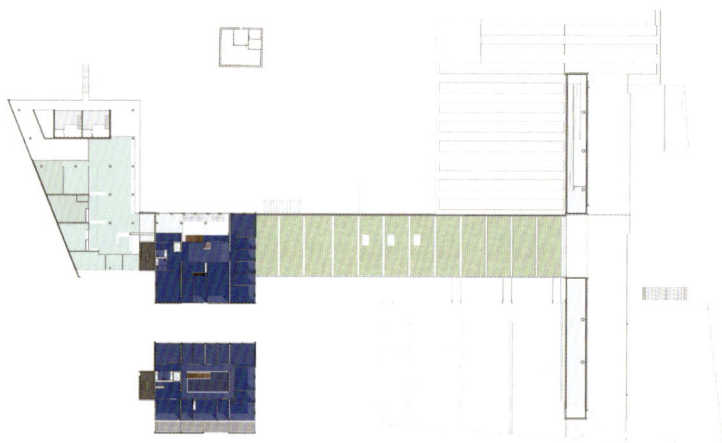

9

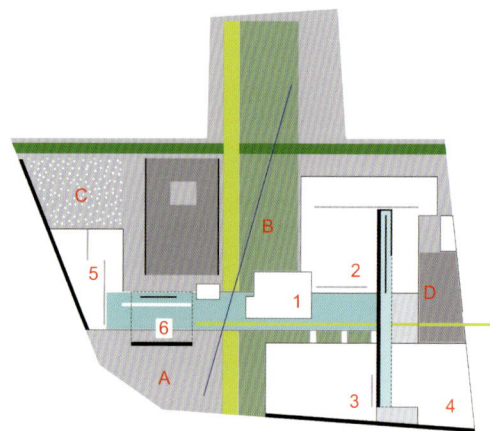

10

13. Hall leading to the exhibition rooms –
installation by Varini
14. Exhibition space section

13. 通往展览室的大厅展有瓦累尼作品
14. 展览厅剖面图

15. Exhibition room with 9 metre ceiling
height
16. Section/perspective of the permanent
exhibition room
17. Section/perspective of the temporary
exhibition room

15. 9米高的展览厅
16. 常设展览室透视剖面图
17. 临时展览室透视剖面图

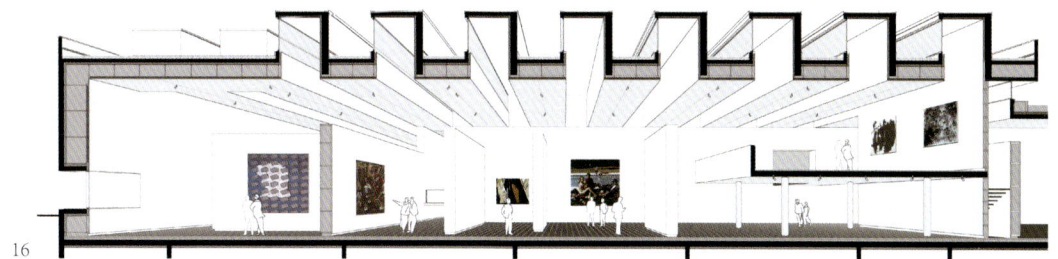

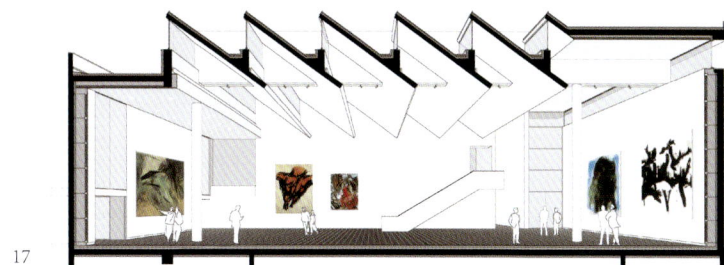

21

18. View of the exhibition rooms
19. Permanent exhibition room
20. Bookshop
21. Permanent exhibition rooms on the mezzanine
22. Temporary exhibition room

18. 展室内景
19. 临时展室
20. 书店
21. 夹层上的常设展览室
22. 临时展览室

22

MALRAUX MUSEUM
马尔罗博物馆

建筑师 Architect: Beaudouin-Husson
地点 Location: Le Havre
完工日期 Completion Date: 1999
摄影师 Photographer: Jean-Marie Monthiers

This museum in Le Havre was one of the pioneering buildings of the early 1960s that marked an evolution in the history of architecture. Its architects, Guy Lagneau, Jean Dimitrijevic and Michel Weill, talented students of Perret, all became successful after its completion. The project's influence was not immediate, and yet the ideas of Lagneau and his engineer Jean Prouvé have continued to exert their influence to this day.

Beaudouin-Husson Architects were commissioned to transform the museum, and did so with the encouragement and help of Guy Lagneau and his partners, Jean Dimitrijevic and Michel Weill. One of the solutions used to preserve the double-height volume was to gather all the peripheral spaces (library, conference room, etc.) in the two first spans of the building, using the untapped space in the eaves. The flexibility of the original design was partly preserved in the spatial relationship between the temporary exhibition room and the large volume of

the permanent collections.

The degree of fluidity is subtly adapted to suit collections that range from modern to old master paintings. Bit by bit one is led into the more intimate spaces. The ceiling follows suit: glazed in the nave and around the outer edges of the building, it becomes opaque above the mezzanine to gain height in the free volumes of the eaves. To filter the natural light, glass sun screens pivot vertically throughout the day in the double skin of the west façade. In the east

façade, the floor stops abruptly to create a double-height space that totally frees up the mezzanne. In this luminous void, a rail 25 metres long, suspended from the structure, seems to float in mid air. These two levels of exhibition spaces are linked by a ramp strengthening and echoing the visual continuity between the museum and the sea.

6

1. Night exterior view
2. General interior view
3. North façade
4. Ground floor
5. Library
6. The ramp

1. 夜间外景
2. 总体内景
3. 北立面
4. 位于一楼的展览室
5. 图书馆
6. 连接上下楼层的坡道

这座位于勒阿弗尔的博物馆是20世纪60年代早期的一座先锋建筑，标志着建筑的进化。它的建筑师盖·拉尼奥、让·迪米特里杰维克和米歇尔·维尔都是佩雷学院的优秀毕业生，在项目完成之后都取得了事业的成功。项目的影响力并没有马上显现出来，但是拉尼奥和他的工程师让·普洛弗的设计理念的影响一直延续至今。

Beaudouin-Husson建筑事务所受委托对博物馆进行改造，他们的工作受到了盖·拉尼奥、让·迪米特里杰维克和米歇尔·维尔的鼓励与支持。为了保护博物馆原有空间的双高结构，建筑师将所有次要空间都设在了建筑底下的两层。利用屋檐下的未利用空间。原始设计的灵活性在临时展厅和大量永久性收藏品的空间关系上得到了体现。

流动性与现代和古典画作巧妙地相称。参观者被逐步引入更为私密的空间。中殿和建筑外围都采用了玻璃装配，中层楼的上方是不透明的，一直上升到檐口自由空间的高度。为了过滤自然光，西立面的玻璃遮阳板会随着日光的变化而垂直转动。东立面的楼面截然而止，打造了双高空间，将中层楼空了出来。在这个明亮的空间中，25米长的围栏从建筑结构上悬吊下来，看似悬浮在半空中。两层的展览空间通过坡道连接起来，加强并回应了博物馆与海洋之间的视觉联系。

7. The ramp
8. Entrance level plan
9. Mezzanine plan
10. Suspended gallery
11. Upper level gallery

7. 坡道
8. 入口层平面图
9. 夹层平面图
10. 悬挂的画廊
11. 上层画廊

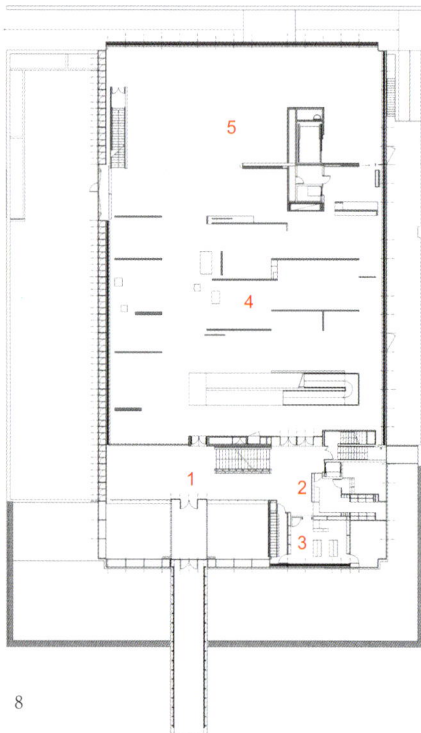

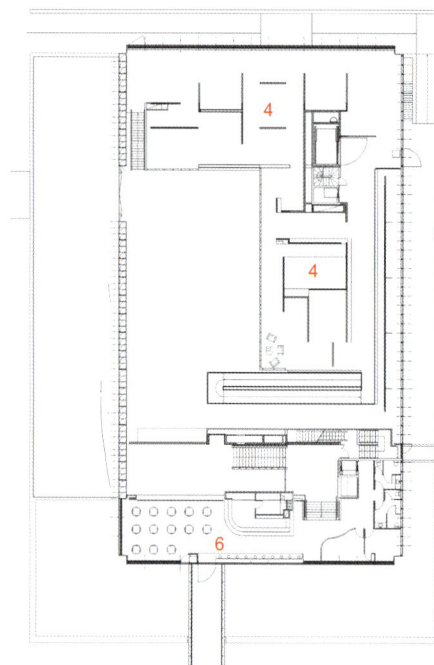

1. Entrance
2. Reception
3. Bookshop
4. Permanent exhibition room
5. Temporary exhibition room
6. Café

1. 入口
2. 前台
3. 书店
4. 常设展览室
5. 临时展览室
6. 咖啡厅

10

11

STRASBOURG MUSEUM OF MODERN AND CONTEMPORARY ART

斯特拉斯堡现当代艺术博物馆

建筑师 Architect: Ateliers AFA
– Adrien Fainsilber & Associés
地点 Location: Strasbourg
完工日期 Completion Date: 1998
摄影师 Photographer: Edith Rodeghiero

Stretching along the River Ill, the Museum of Modern and Contemporary Art (MAMC), embraces Strasbourg and reconstructs the urban landscape through its link with several strong features: the river, the Saint-Jean commandery, the Vauban dam and the administrative headquarters of the Bas-Rhin department on the opposite bank. Its relationship with the water, light and the historic city have strongly influenced the organisation of the site.

The museum is spread over three floors, one of which is partially a mezzanine. The wide central nave forms the backbone of the museum, hosting the welcome desks and opening onto the different public areas. As a nod to tradition, the contrast in materials and textures – glass, pink granite and smooth, white concrete resin panels – evokes traditional Strasbourg architecture with its white plaster surfaces outlined in pink sandstone.

Adjacent to the entrance on the ground floor, the modern art room and a gallery housing the works of Gustave Doré run the length of the riverside wing. The modern art room devoted to the sculptures of Jean Arp is the showpiece of the museum. The Doré gallery fits the works that it houses. It is 12 metres high and lit by a skylight, with a lower ceilinged area for drawings and sculptures. On the opposite side of the nave, a temporary

exhibition room is both high ceilinged and adaptable through its moveable partitions, so it can respond to the most unexpected demands of contemporary. On the first floor, four exhibition rooms fill the length of the building. And on the roof, with the restaurant, Mimmo Paladino's horse sculpture can be seen from afar, heralding the presence of the museum in the city.

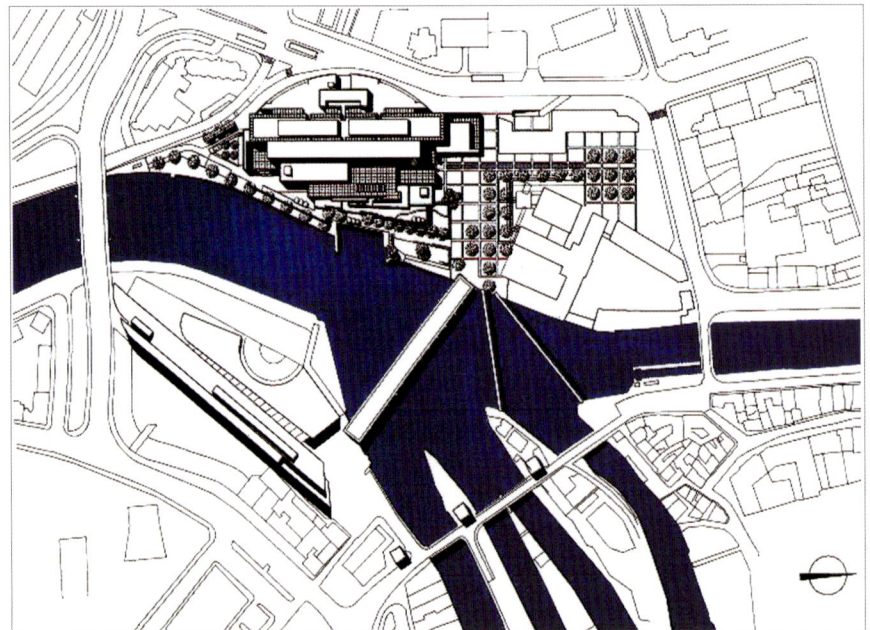

1. View from the Ill
2. Exterior view
3. The glass façade
4. Master plan
5. Panoramic terrace
6. Entrance detail

1. 临伊尔河外景
2. 博物馆外景
3. 玻璃立面
4. 总规划图
5. 全景露台
6. 入口细部

斯特拉斯堡现代艺术博物馆沿着伊尔河延伸，与城市的若干个突出的特征相互联系：伊尔河、圣让会所、沃邦水坝和河对岸的下莱茵区行政总部，并以此重塑了城市景观。它与水、光和古城的关系对场地的组织结构有着强烈的影响。

博物馆共有三层，其中一层的一部分是中层楼。宽阔的中庭形成了博物馆的支柱，立面设置着迎宾台，并通往不同的公共区。玻璃、粉花岗岩、光滑白水泥树脂板等材料和材质的对比与传统斯特拉斯堡建筑的白石膏表面和粉砂岩结构遥相呼应。

紧邻一楼入口的现代艺术展室和画廊内收藏着古斯塔夫·多雷的作品，横跨河畔的一侧。展出让·阿尔普雕塑作品的现代艺术展室是博物馆的招牌。多雷画廊与其所展示的作品十分相称，高12米，采用天窗照明，较低的吊顶区域则用来展示画作和雕塑。中庭另一侧的临时展厅同样采用高吊顶，移动屏风可以灵活地对空间进行布局，满足当代艺术品展示的需求。二楼共有四个展室，与楼体等长。人们从远处就能看到博物馆屋顶密莫·帕拉迪诺的骏马雕塑，彰显了博物馆的存在感。

7. Escalator and footbridge in the nave
8. Detail of the staircase
9. Cross section
10. Nave interior
11. Framed view of La Petite France

7. 正厅人行桥和电梯
8. 楼梯细部
9. 横截面
10. 正厅内部
11. 小法兰西景点成为博物馆的框景

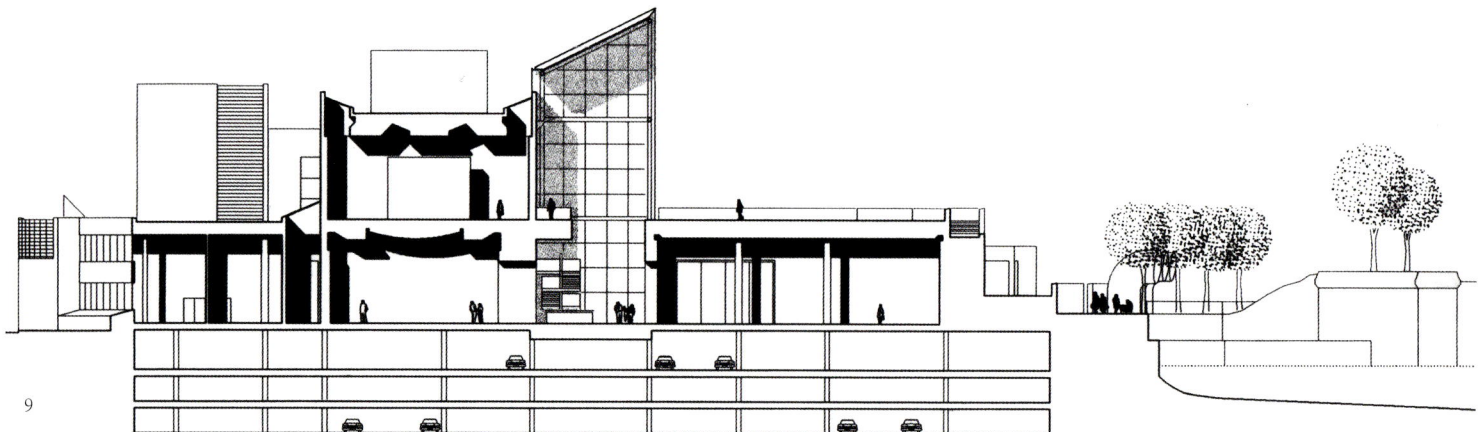

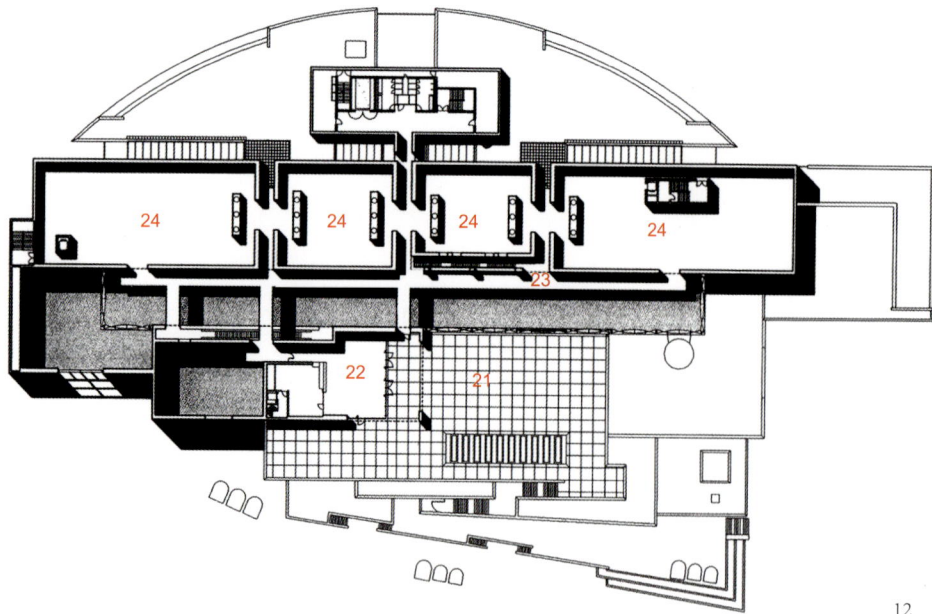

12

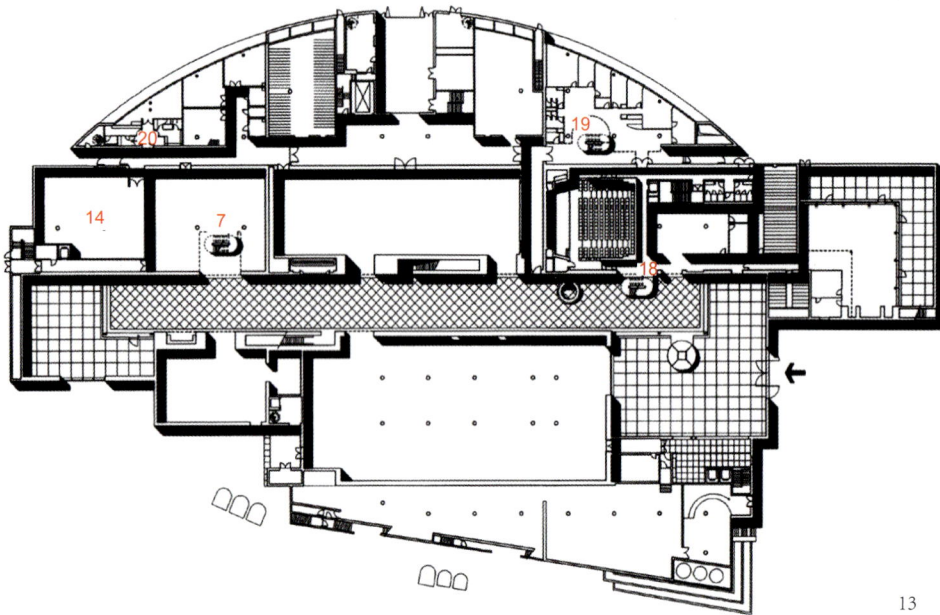

13

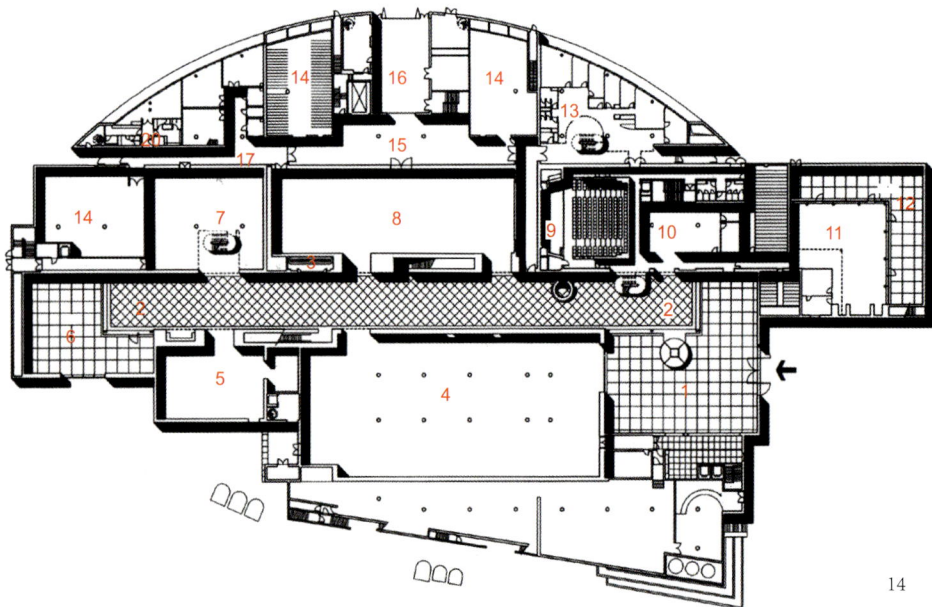

14

16. The nave and the corridor
17. Exhibition room
18. Detail of the administration staircase

16. 正厅与二楼的连廊
17. 展览室
18. 行政办公室楼梯细部

QUAI BRANLY MUSEUM
布朗利河岸博物馆

建筑师 Architect: Jean Nouvel,
with Gilles Clément (Landscape),
Patrick Blanc (Vertical garden),
Yann Kersalé (Light installation)
地点 Location: Paris
完工日期 Completion Date: 2006
摄影师 Photographer: Philippe Ruault
(pp.274-279), Roland Halbe (pp.281-283)

The Quai Branly Museum is a meditation on the other and elsewhere. The space that houses the objects is neither religious nor sacred. Jean Nouvel wanted to surround them with mystery, to give them back their own life. Rejecting western rationality, he preferred to create them a home where they would be together, conversing with each other, or withdrawn into their existence alone. Instead of a clear museum approach, which shines light on the works in white and neutral rooms, in glass cases, he preferred appearance and disappearance. No magic or illusion: Jean Nouvel opted for allusion.

A single floor was proposed, an open gallery. The collections are grouped by continent, but the geography that results is an imaginary map, a long gallery measuring more than 200 metres. To welcome the objects it was necessary to lose one's bearings, to disorientate. This place is the foyer of a home. Half-light dominates, necessary for conservation and for this work of adaptation. The

light is diffuse, through latticework. The materials are soft, tactile, ductile, without impediments on their surfaces. Motifs appear at the same time as they hide. The floors are not uniform or flat: walking, balance, are in play to accompany the exploration. The colours are warm, dark, deep and absorb the light.

Bearings are needed, however: long and narrow, the gallery is between two façades that the daylight makes vibrate differently. On the Seine side, the motifs of forests

4

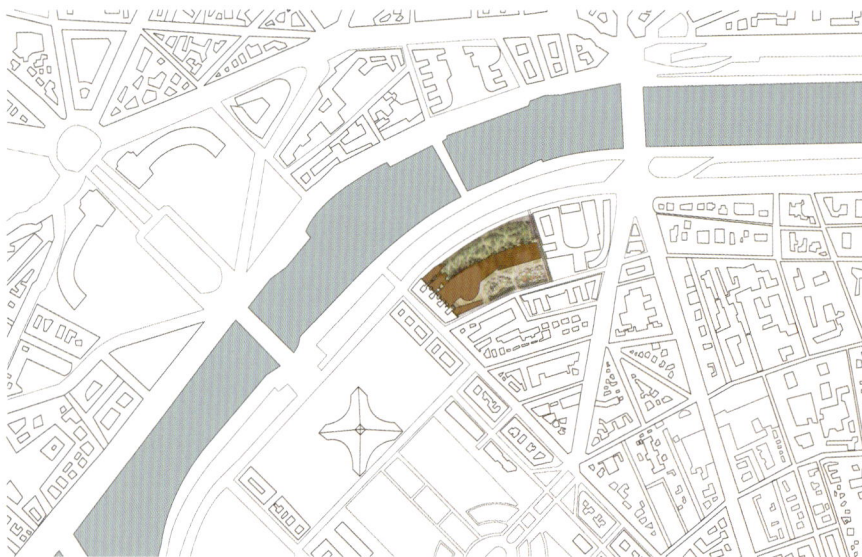

5

soften the north light (a film in the thickness of the glass panes reflects it towards the exterior) and on the building housing the administration a façade was designed with the botanist Patrick Blanc. On the south side, very filtering glass is behind panels that play like the slats of a blind. A geometric break allows for the framing of the Eiffel Tower. Thus, the museum is not enclosed within its walls, but enshrined by the thickness of its façades that filter its relationship with the outside world.

6

1. Façade on Quai Branly
2. Branly and Auvent buildings, administration
3. North façade, garden
4. South façade of the museum
5. Master plan
6. Collections floor

1. 面对布朗利河岸的博物馆立面
2. 作为行政用途的布朗利楼和欧旺楼
3. 博物馆北立面和花园
4. 博物馆南立面
5. 基地位置图
6. 展览厅

布朗利河岸博物馆是对别处的沉思，展示着藏品的空间没有丝毫的宗教气息。让·努维尔希望赋予它们神秘的氛围，使它们回归自我。他拒绝了西方的理性主义，更倾向于为藏品打造一个能够聚集在一起、相互交流或是独自隐匿的家。传统的博物馆通常将藏品放置在白色或中性的房间中，罩上玻璃，在灯光下清晰地展示；努维尔则倾向于若隐若现。这里没有魔法或幻觉，努维尔选择了暗示。

建筑采用单层开放式展厅结构。藏品根据大洲分类，地理位置所描画出的虚构地图绵延200多米。为了欣赏藏品，人们会在展厅中迷失方向。博物馆就像家的门厅，暗光适合会话和展示藏品。透过栅格，灯光四处弥散。

设计所选用的材料柔和、有质感、可塑性强，在表面上没有丝毫障碍。各色图案若隐若现。楼面呈现出不同的特色：漫步，平衡，一切都伴随着探索的流程。设计色彩温和、昏暗、深邃，吸收了光线。

然而，方位感也是必须的：或长或短，夹在两面外墙之间的走廊随着日光摇摆不定。塞纳河一侧，森林图案柔化了北部光（一种嵌入玻璃板的薄膜，向外反射）；行政楼的一面外墙则由建筑师与植物学家帕特里克·布兰克一起设计。南侧的滤光玻璃装配在百叶窗式面板上。透过几何造型的缝隙依稀可以看到埃菲尔铁塔的轮廓。这样一来，博物馆并没有被封闭在墙内，而是潜在外立面之间，保持着与外界的联系。

7. Collections floor, north façade
8-9. Collections floor, south façade
10. Collections floor
11. Ground floor plan

7. 展览厅，北立面内部
8、9. 展览厅，南立面内部
10. 展览厅
11. 一楼平面图

1. Entrance hall
2. Ticket desks
3. Temporary exhibitions
4. Administration
5. Kerchache room
6. Bookshop
7. Cafeteria

1. 入口大厅
2. 售票处
3. 临时性展览空间
4. 行政空间
5. Kerchache 阅览室
6. 书店
7. 咖啡厅

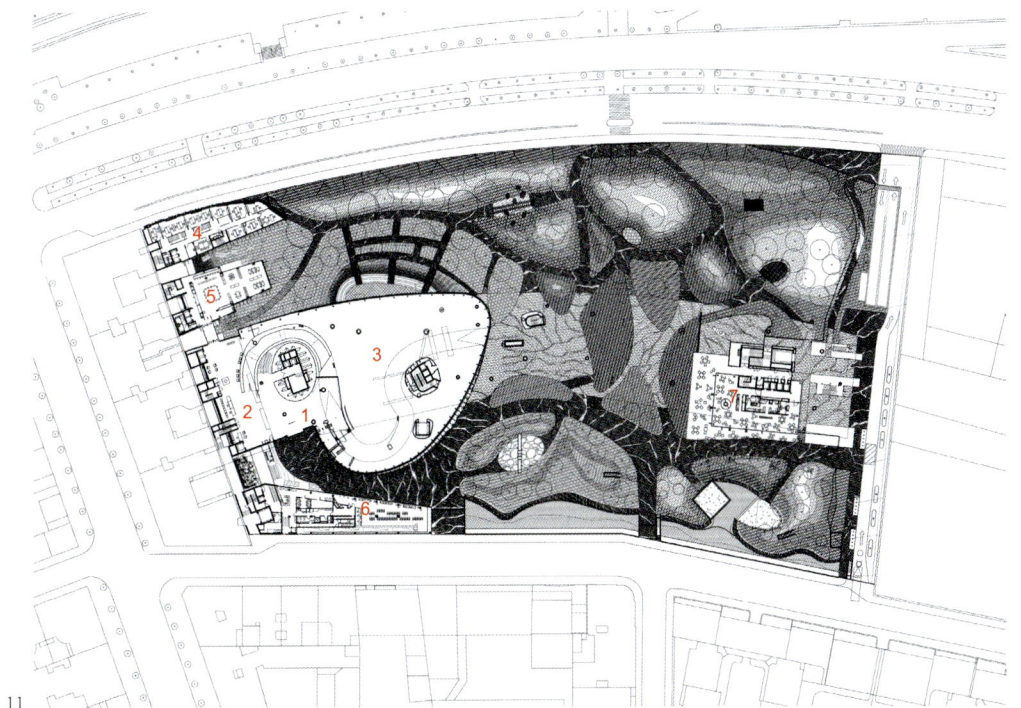

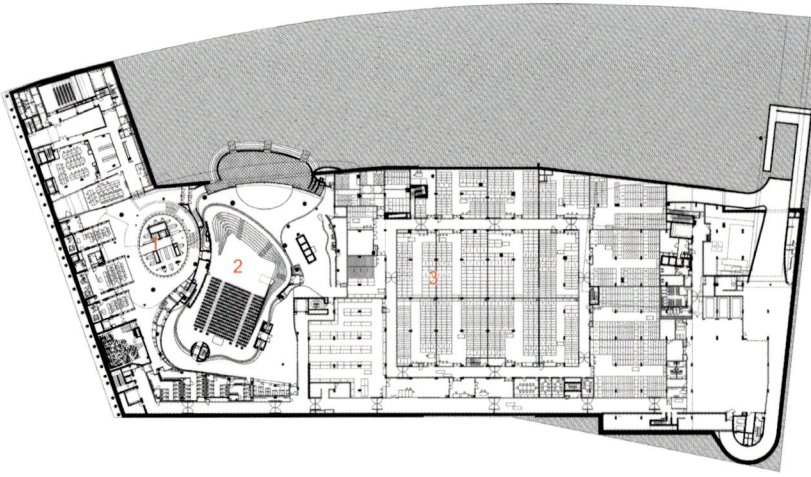

12

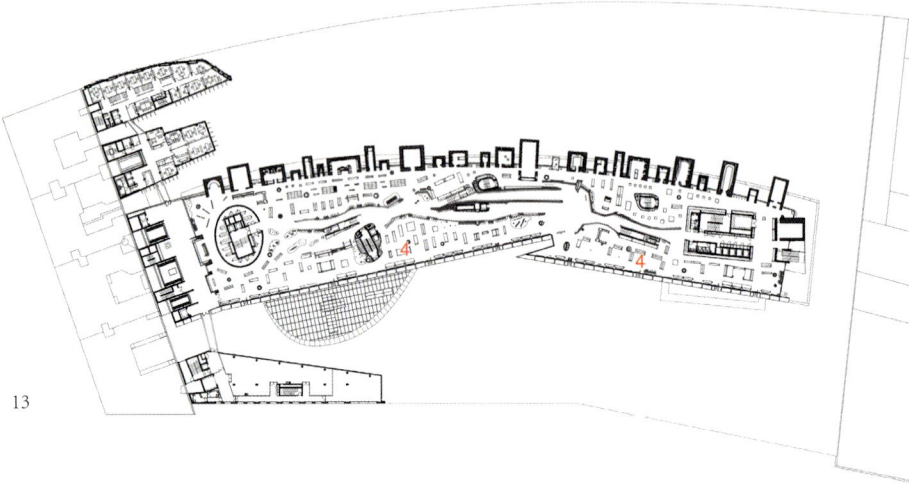

13

1. Instruments tower
2. Auditorium
3. Reserves
4. Collections floor
5. Mezzanines
6. Terrace
7. Restaurant
8. Mediatheque

1. 乐器之塔
2. 表演厅
3. 储藏空间
4. 收藏品展览厅
5. 展览厅夹层
6. 屋顶平台
7. 餐厅
8. 多媒体图书馆

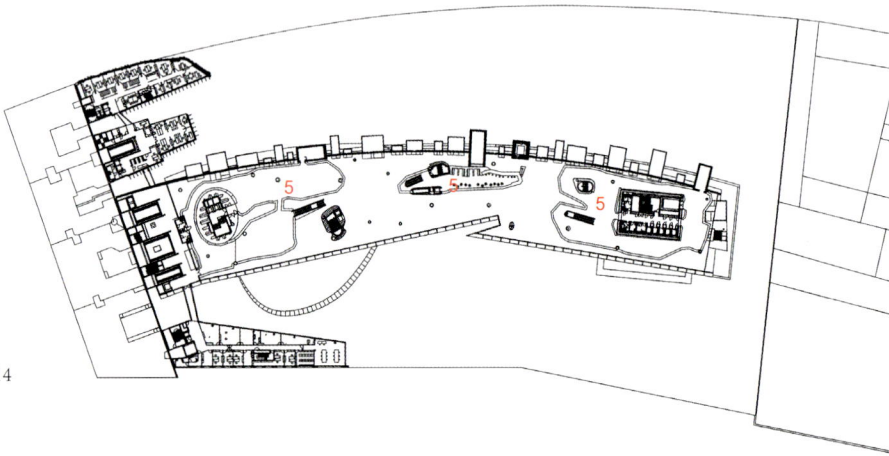

14

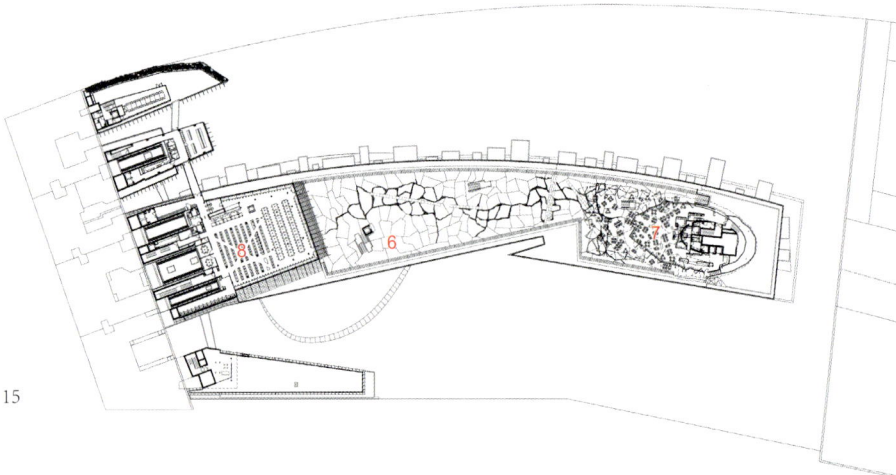

15

17

18

17. Collections floor
18. Longitudinal section of the museum (BB)
19. Collections floor, the serpent
20. Cross section of the Branly, Auvent and Université buildings (AA)
21. Cross section of the restaurant, museum and car park (CC)
22. Cross section of the atrium, auditorium, ramp and museum (DD)

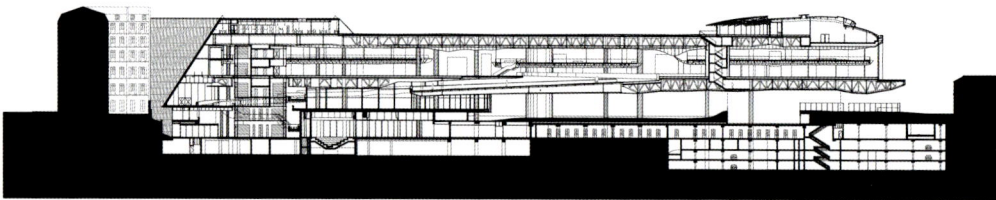

17. 收藏品展示厅
18. 长向剖面图 (BB)
19. 收藏品展示厅，蛇形般的空间场景
20. 横向剖面图 (AA)：布朗利、欧旺和大学等建筑楼
21. 横向剖面图 (CC)：餐厅、展览厅与停车场
22. 横向剖面图 (DD)：中庭、表演厅、斜坡与展览厅

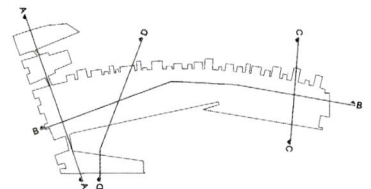

19

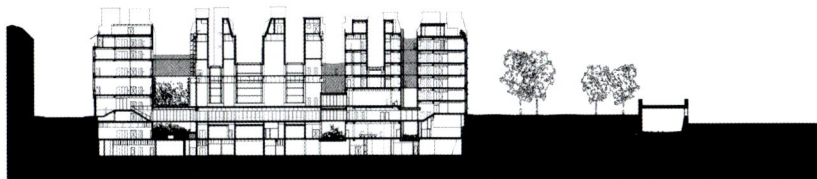

20

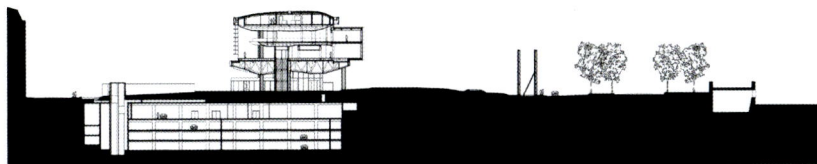

21

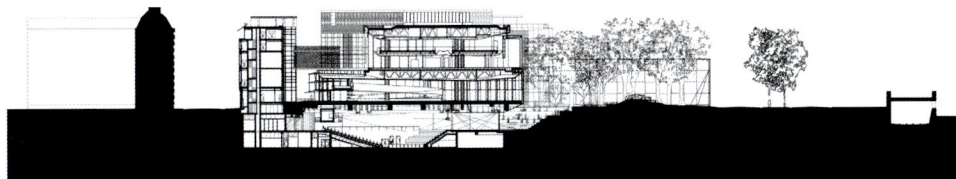

22

WÜRTH MUSEUM
伍尔特博物馆

建筑师 Architect: Jacques & Clément Vergély
地点 Location: Erstein
完工日期 Completion Date: 2008
摄影师 Photographer: Erick Saillet

The Reinhold Würth collection was to be housed on the same industrial site as the Würth-France headquarters. To express the extreme variety of the collection and its contrast with the functionality of the head office, a diversification and neutrality of the spaces was called for. Two perfectly identical, parallel parallelepipeds separated by a welcome space form the morphology of the building.

The choice of rough concrete was obvious. The two monoliths had to express the unadorned abstraction of the two built bodies and their protecting symbolism through the use of a single material. The proximity of the head office imposed a confrontation that enhances both buildings, the glass façades of the head office finding a pertinent counterpoint in the opacity and courseness of the rough concrete. The functional sequences are very easy to read. In the east all the services offered to the public are gathered on two levels,

1

while the west houses the delivery and maintenance areas, and the reserves. In the centre are the rooms devoted to artistic expression. In the north an exterior, uncovered space opens up, in which sculptures may be placed.

Separated by the music rooms, the exhibition spaces have deliberately contrasting dimensions. The three rooms have common characteristics in terms of materials: a polished concrete floor, walls made up of wood and plaster

2

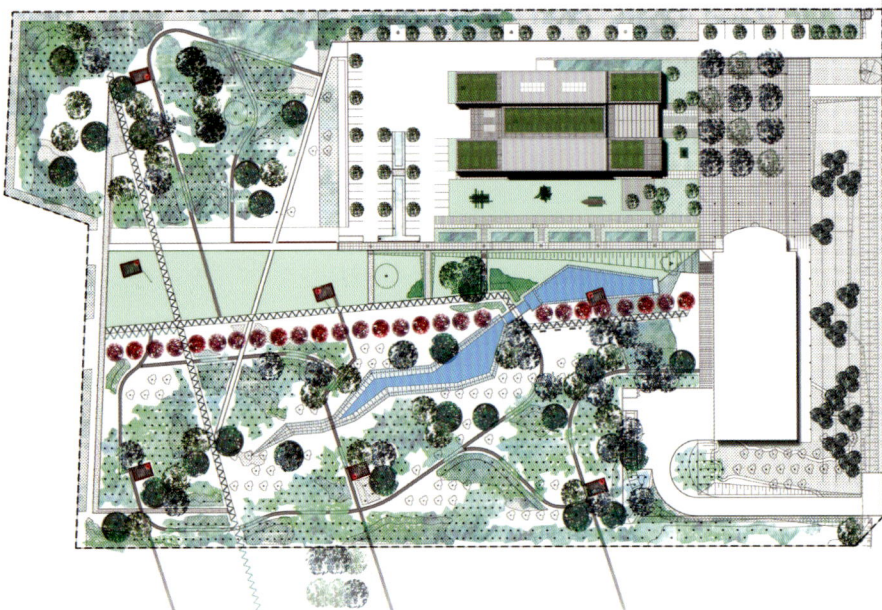

combinations that allow for easy hanging fittings, a false ceiling integrating the air conditioning system. The walls and the ceiling are white. The major common characteristic of these exhibition spaces resides, in the end, in their unusual use of natural light.

1. South façade of the museum from the Würth France head office building
2. South façade of the museum and the park
3-4. South façade of the museum
5. Master plan
6. South façade of the museum, the covered gallery and the park – play of reflections

1. 临法国伍尔特总部的博物馆南立面
2. 博物馆的南立面和公园
3、4. 博物馆南立面
5. 总规划图
6. 博物馆南立面、有顶长廊和公园的河边倒影

莱茵霍尔德·伍尔特的收藏品都被安置在法国伍尔特总部的同一个工业场所。为了展现藏品的多样性并且使收藏空间与总部办公楼的功能区分开，急需一个多样化的中立空间。最终，博物馆建筑由两座完全相同的并立平行六面体组成，二者由迎宾区相连，形成统一的整体。

毛面混凝土的选择十分显眼。两个楼梯结构必须表现出朴素抽象的形象，运用单一的材料保护它的象征意义。博物馆与总部大楼相邻，总部的玻璃外墙与毛面混凝土的不透明和粗糙感形成了鲜明对比。博物馆的功能区设置十分清晰：所有向公众提供的服务设施都聚集在东侧的两层楼，西侧则设置着运送维护区和仓库。中央的空间全部是艺术展厅。建筑北侧的露天空间内可以用来放置雕塑作品。

展览区与音乐厅隔开，具有刻意相反的设计。三间展厅采用了相同的材料：抛光水泥地面、木板和石膏混合墙面以及配置空调系统的假吊顶。墙壁和天花板都是白色的。展览厅的主要共同点在于它们对自然光的独特运用。

MUSÉE **WÜRTH** FRANCE ERSTEIN

9

7. Museum entrance, east façade overlooking the forecourt, night view
8. South façade of the museum and the park, night view
9. Forecourt with a sculpture by Bernard Venet, 220° Arc x 5, 2002, Collection Würth, Inv. 10552
10. Cross section

7. 夜晚的博物馆入口和东立面
8. 夜晚的博物馆南立面和公园
9. 前庭广场上的伯纳德·维内特雕塑作品，*220° Arc x 5*，2002年，伍尔特博物馆收藏，编号10552
10. 横断面

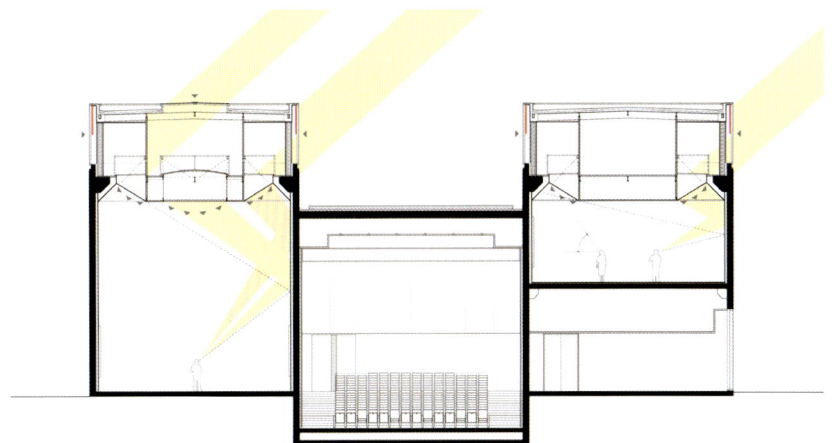

10

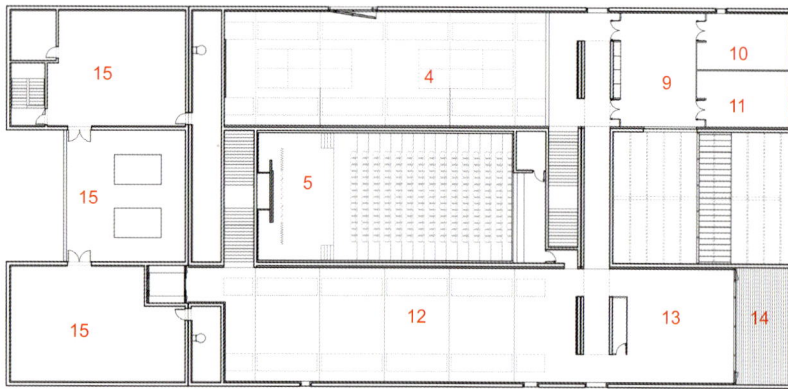

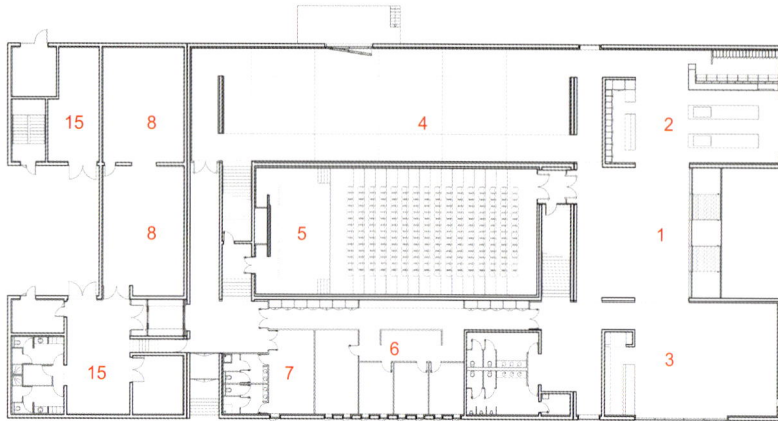

11

12

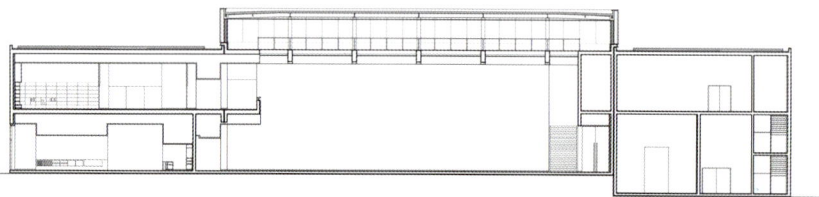

13

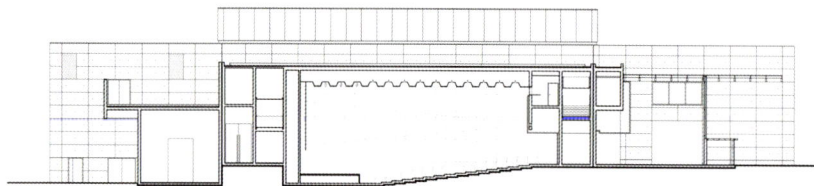

14

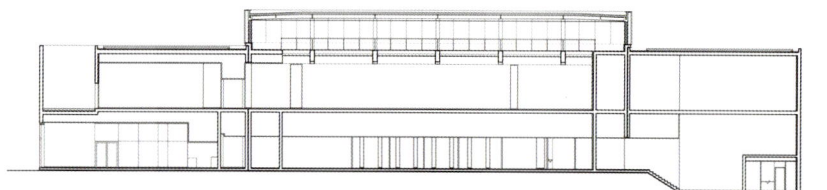

15

16

17

18

19

GESTES PLASTIQUES

La concentration du geste comme principe fondamental de la création artistique constitue le thème de cette dernière partie de l'exposition où sont rassemblés des partis pris aussi différents que ceux de Hans Hartung, Lucio Fontana, Günther Uecker ou Raimund Girke. Tous témoignent de l'association entre spontanéité inspirée et savoir-faire calculé qui, grâce au geste de l'artiste, se transforme en expression de la virtuosité créatrice.

PLASTISCHE GESTEN

Die konzentrierte Geste als Grundprinzip künstlerischer Kreation steht im Mittelpunkt des fünften und letzten Teils der Ausstellung, die so unterschiedliche Positionen wie Hans Hartung, Lucio Fontana, Günther Uecker und Raimund Girke zusammenführt. Ihnen allen ist die perfekte Mischung aus inspirierter Bewegung und kalkuliertem Wissen eigen, die, in einer plastischen Geste konkretisiert, zum Ausdruck schöpferischer Virtuosität wird.

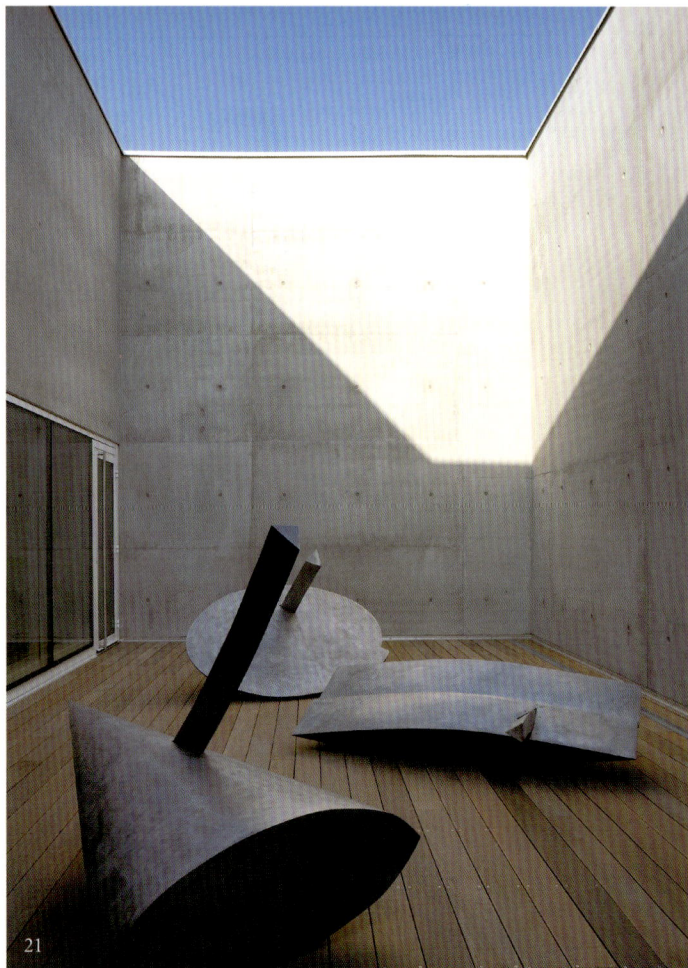

20. South small exhibition room and patio
21. Patio
22. North main exhibition room
23. North main exhibition room and mezzanine
24. Museum auditorium

20. 小型南展厅和中庭
21. 中庭
22. 北主展厅
23. 北主展厅和夹层
24. 博物馆演映厅

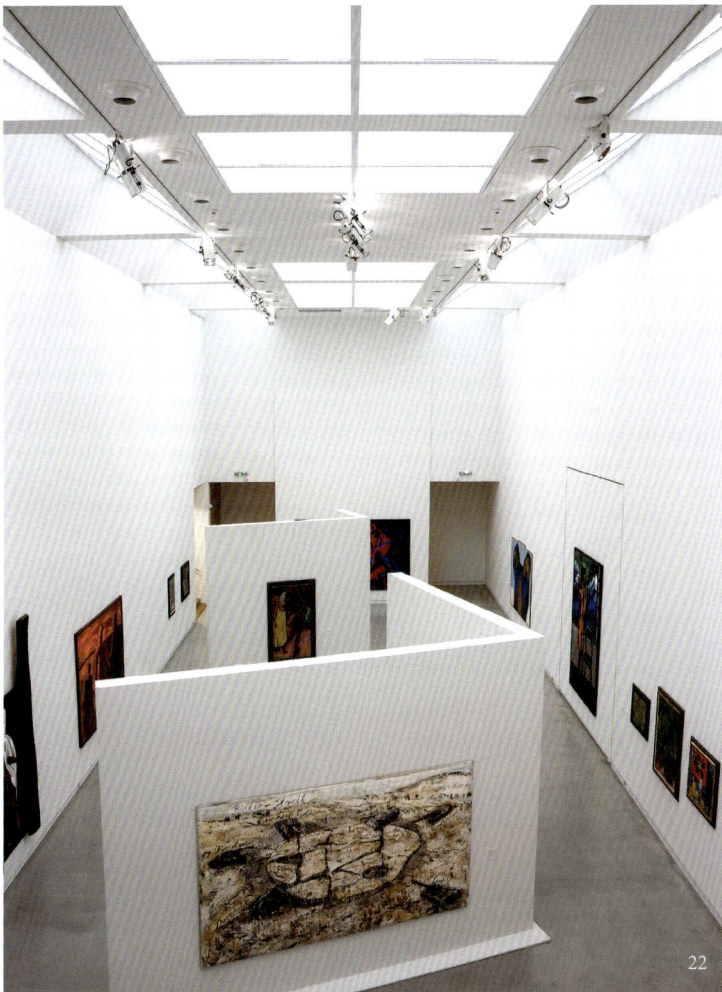

22

EXPRESSION-FIGURATION

23

24

VESUNNA GALLO-ROMAN SITE-MUSEUM
韦索纳高卢罗马遗址博物馆

建筑师 Architect: Jean Nouvel
地点 Location: Périgueux
完工日期 Completion Date: 2003
摄影师 Photographer: Philippe Ruault

In Périgueux, the site-museum designed by Jean Nouvel preserves and exhibits the remains of a large Gallo-Roman residence, the Vesona mansion (Domus de Vésone). The daily life of ancient times is presented through archaeological collections going back to the 1st to 3rd centuries A.D. Jean Nouvel introduced the project in this way: "We must reveal and protect this site. With nobility. With tact. With clarity, using the sensitivity and culture of our age, through an architecture that will have the awe-inspiring privilege of coming face to face with the ghosts of antiquity."

The visitor first enters a long building known as the "thick wall" running along the west side of the site and forming a screen that directs the gaze towards the Vesona Tower. This building groups the reception, a shop, one section of

the "museum" circuit and the technical services. Two floors designed as mezzanines overlook the remains of the mansion.

The museum's main attraction, the mansion reveals the scope of the remains of the ancient city. Everything is organised around it. The heart of the mansion, with the garden, a peristyle and the rooms arranged off it, is protected by

a "large sheltered courtyard" covered with a huge metal umbrella held up by 9-metre pillars. The covering juts out several metres, protecting the museum from direct sunlight. Inside, the visitor can walk among the remains on wooden decking placed on a metallic structure.

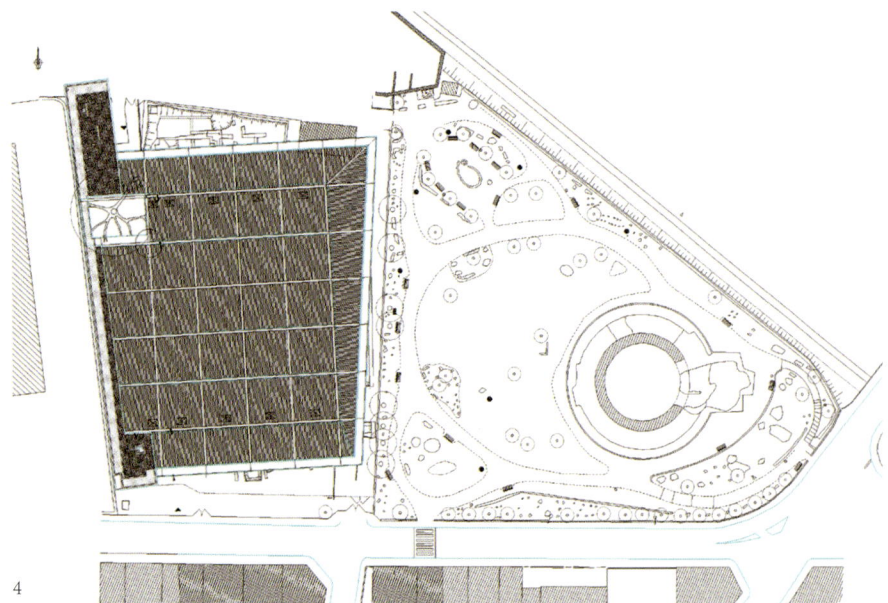

5

1. North façade overlooking the park
2. North façade at night
3. Museum entrance
4. Master plan
5. North façade, M. Taillefer's house

1. 面向公园的北立面
2. 北立面夜景
3. 博物馆入口
4. 总体配置图
5. 北立面,底处为Taillefer先生之家

在佩里格,由让·努维尔设计的遗址博物馆保存并展示了一个大型高卢罗马住宅——韦索纳庄园。可追溯到公元1世纪到3世纪的考古文物呈现了远古时代的日常生活。让·努维尔以这种方式来介绍项目:"我们必须展示并保护这一遗址,以一种崇高、巧妙、清晰的方式,利用我们时代的敏感性和文化,通过建筑来与文物的灵魂面对面。"

参观者首先进入的是以"厚壁"著称的长楼,它沿着遗址西侧延伸,犹如一扇屏风,将人们的视线导向韦索纳塔。长楼内汇集了前台、商店、博物馆展览路线的一部分和技术服务设施。两层中层楼俯瞰着庄园的遗迹。

博物馆的主要景点——庄园展现了古代城市的遗迹。所有设施都围绕着它展开。庄园(包括花园、列柱围廊和周边的房间)的中央受到了一个"大型带顶庭院"的保护,上方是一把由9米高立柱支撑的金属伞。金属屋顶向外延伸了数米,保护博物馆不受阳光直射。在博物馆内部,来访的游客可以行走在由金属结构支撑的木平台上进行参观。

6. Est façade, view into the museum's interior with the mezzanines in the background
7. Domus floor plan
8-9. Museum, Domus floor

6. 从东立面看博物馆内部，底处为带有两层中层楼的长廊
7. 韦索纳庄园平面图
8、9. 博物馆结构和庄园遗迹

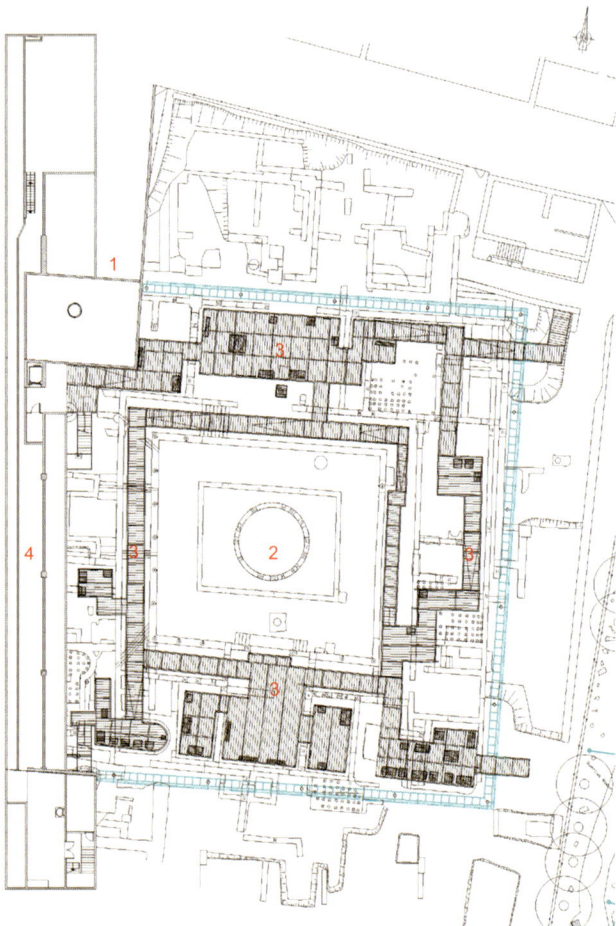

1. Entrance
2. The domus, remains of a large Gallo-Roman residence
3. Wooden decking placed on a metallic structure
4. The "thick wall", mezzanines grouping the reception, a shop, one section of the "museum" circuit and the technical services

1. 博物馆入口
2. 韦索纳庄园，一个广大高卢罗马遗迹的一部分
3. 由金属结构支撑的木质平台
4. 被称作"厚壁"的长楼，以两层中层楼的空间汇集了前台、商店、博物馆展览路线的一部分和技术服务设施

10-12. Museum, Domus floor

10-12. 博物馆结构和庄园遗迹

BLIESBRUCK-REINHEIM GALLO-ROMAN BATHS

布里斯布鲁克—莱茵汉姆高卢罗马浴场

建筑师 Architect: Frédéric Jung
地点 Location: Bliesbruck
完工日期 Completion Date: 1993
摄影师 Photographer: Hervé Abbadie

This museum-style project for covering the Gallo-Roman Baths forms a first stage in a larger scheme for protecting the archaeological digs of the European Archaeological Park of Bliesbruck-Reinheim. The construction avoids any restitution or dangerous pastiche. No part of the structure is planted within the confines of the remains.

It merely signifies the former solidity of this major ancient edifice, bordering one of the city's most important public squares, through the positioning of the porticos supporting this vast copper covering. It is a "museographic machinery" that reveals the site and allows us to read the remains. The museography simply indicates "how to look". The objective is to provide the means of deciphering the different significations of the ruins and to understand the aims and methods

of archaeology. The museum signs are designed to help the visitor interpret what he sees and feels, by providing information that stimulates observation.

The idea is to encourage visitors to walk around the site via a series of footbridges raised above the general levelling of the remains, at approximately the height of the original floors, in order to remove the ambiguity of accessing the

1. North façade
2. West façade
3. Side view of the auditorium
4. South façade
5. General plan

1. 北立面
2. 西立面
3. 礼堂侧面图
4. 南立面
5. 总平面图

remains at the level of the basements and hypocausts and to give a synthetic vision of the arrangement of the rooms and their functioning. This approach naturally protects the remains. The internal museography offers suspended, light and mobile partitions (Venetian-blind style screens) suggesting how the spaces would have been dispersed in antiquity according to today's archaeological theory, which may yet be contradicted by work still to come.

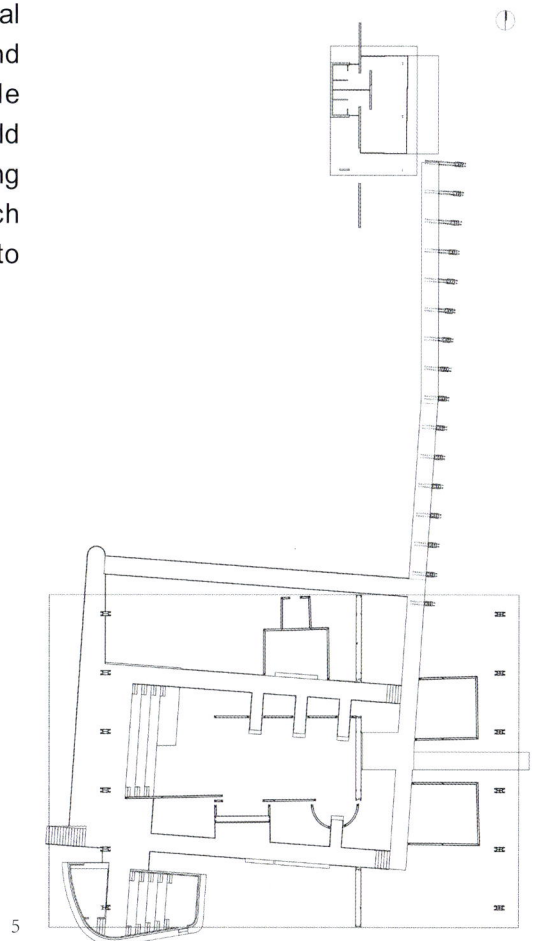

这个博物馆的项目将高卢罗马浴场纳入其中，形成了布里斯布鲁克-莱茵汉姆考古公园考古挖掘现场大型保护计划的第一阶段。项目的建造排除任何大胆修复或仿古的举动，同时，任何建筑结构都不能被植入到遗迹的范围之中。

这座宏伟的古建筑坐落在城市最重要的广场边缘，设计利用门廊来支撑宽大的铜制顶棚。新建筑仿佛一座"博物馆机器"，展示着场地，让我们能够解读遗迹。博物馆的场景设计仅以简单的方式来指示人们"如何观看"。项目的目标是提供解读遗迹意义的方法并让人们理解考古学的目的和方式。博物馆的引导标示通过提供相应的信息来帮助参观者理解他们在现场所看到的景象与感受到的氛围。

设计旨在鼓励参观者通过一系列架高的人行桥（高于遗迹的水平面，约为原来楼面的高度）来环绕场地走动，以避免他们从遗迹地下层或热坑进入产生的混淆，同时也提供了空间布局和功能的整体感。这种方式很自然地保护了遗迹。博物馆的内部设置了轻盈、可移动的悬吊式隔幕（类似威尼斯式百叶屏风），根据当今的考古学理论来展示古时候的空间分割。这些理论有可能被未来的研究所推翻，此种弹性的隔间形式便可以被调整，以适应新的空间分割。

6. Remains of the latrines
7. Approach bridge
8. System of viewing platforms suspended above the Gallo-Roman remains

6. 厕所的遗址
7. 引桥
8. 高卢罗马遗址上方的悬吊观景台

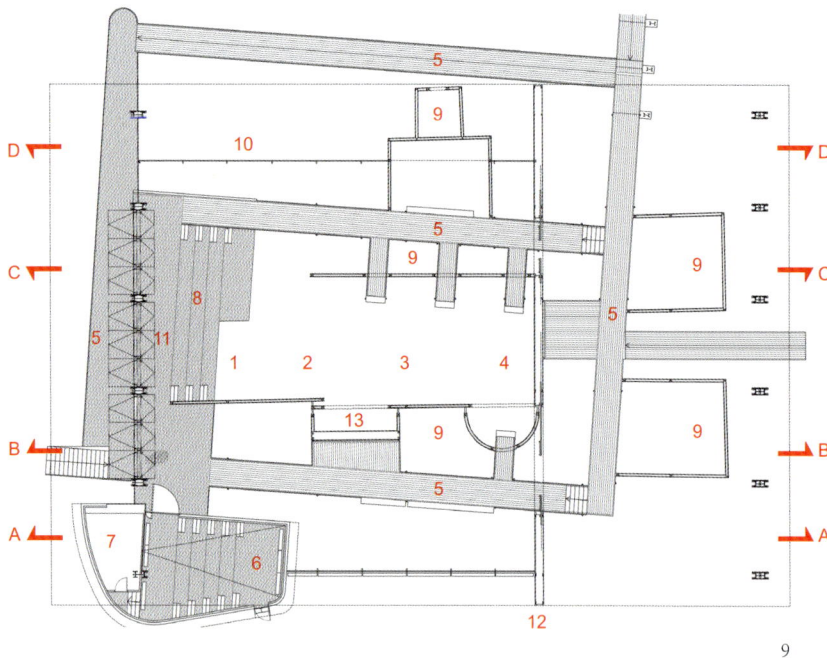

Archeological remains:
1. Ovens and baths
2. Caldarium (hot bath)
3. Tepidarium (warm bath)
4. Frigidarium (cold bath)

Suspended museographical system:
5. Footbridges
6. Auditorium
7. Control room
8. Platforms overlooking the baths
9. Suspended screens (volumetric reconstruction)
10. Protective glass screen
11. Mobile glass screen
12. Thick wall integrating sliding doors (between the public spaces and public facilities)
13. Display case

考古遗迹：
1. 烤箱和浴缸
2. 热水浴
3. 温水浴
4. 冷水浴

博物馆悬吊技术：
5. 人行桥
6. 礼堂
7. 控制室
8. 能够望见浴池的平台
9. 悬吊的屏幕（容积重建）
10. 防护玻璃屏
11. 移动玻璃屏
12. 厚壁整合滑动门（公共场所和公共设施之间）
13. 展示橱窗

9

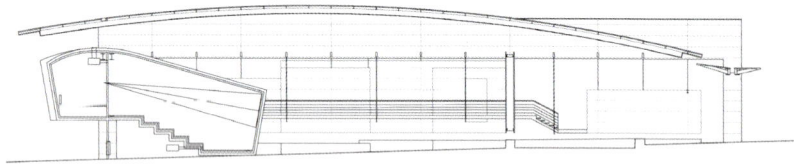

10

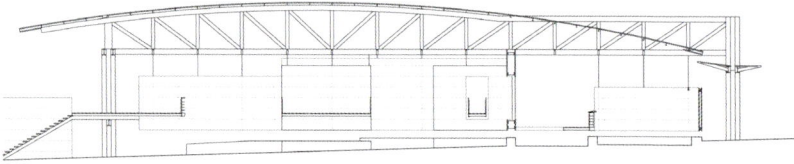

11

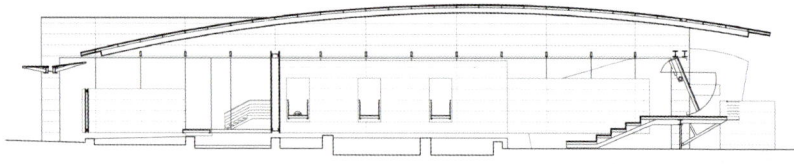

12

9. Plan
10. Section AA
11. Section BB
12. Section CC
13. Section DD
14. Viewing platform over the frigidarium

9. 平面图
10. AA剖面图
11. BB剖面图
12. CC剖面图
13. DD剖面图
14. 冷水浴室

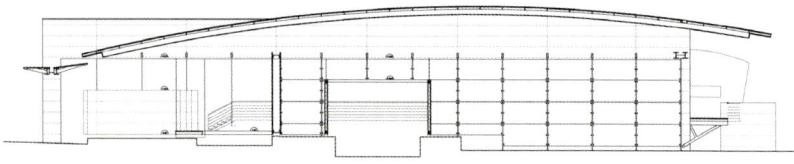

13

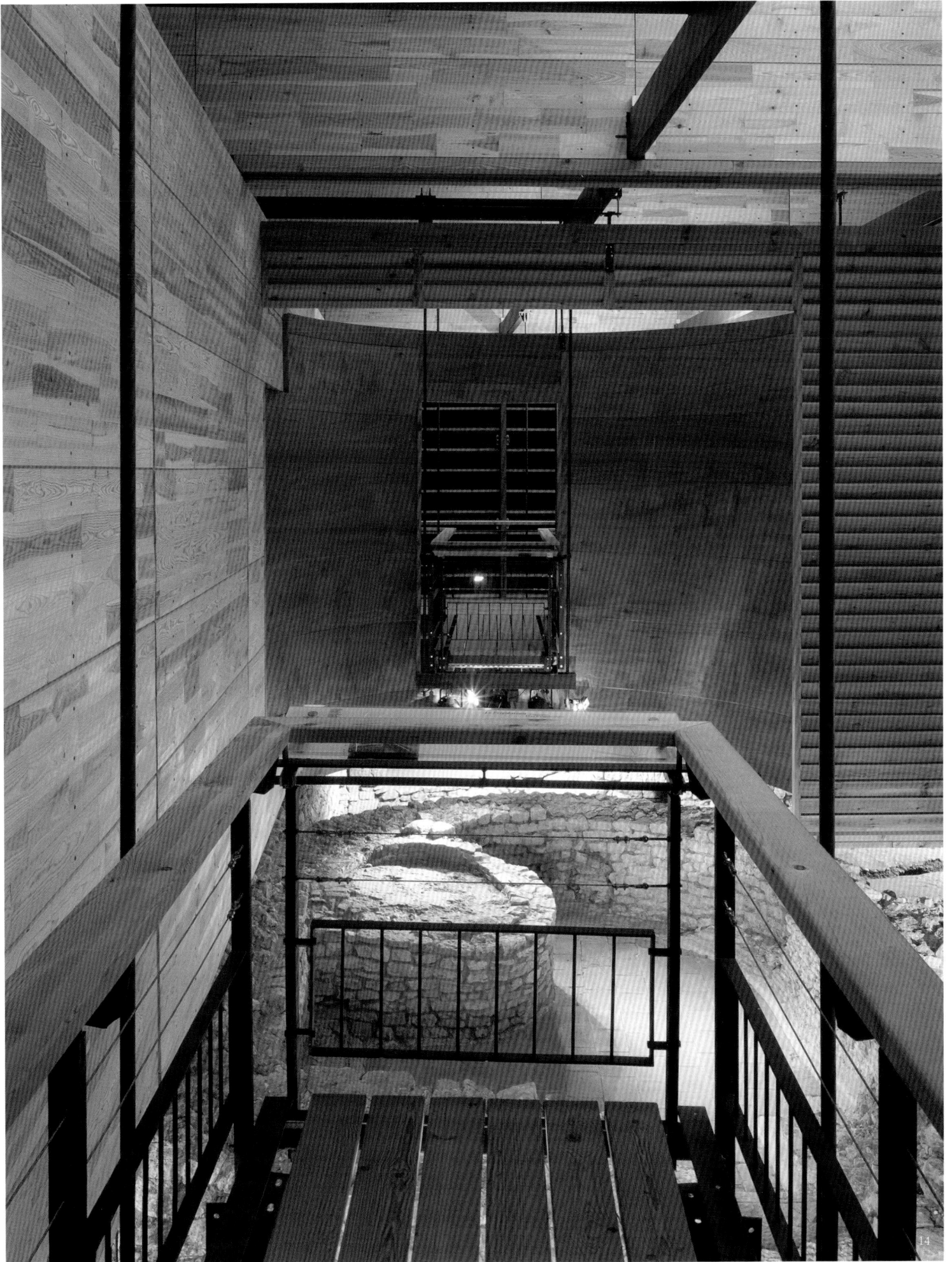

FUTURE
PROJECTS
未来项目

CONFLUENCES MUSEUM

汇流博物馆

建筑师 Architect: Coop Himmelb(l)au
地点 Location: Lyon
完工日期 Completion Date: 2014
图面资料 Visual documents: Coop Himmelb(l)au

On the southern tip of the Presqu'île of Lyon, at the confluence of the Saône and the Rhône, the Confluences Museum is a museum of science and of society. The questions of the future will be decided through the transitional fields of technology, biology and ethics, which form the central theme of the museum.

Conveying the knowledge of our age in a direct and interactive way, it is not only a museum but also a resource for the city. The architecture is both functional and ethereal, creating a hybrid of a museum and urban leisure space, a place that combines education and relaxation, and a connection between the city and culture. Partly inspired by its site, providing an interface between the river and the city, it is composed of two complexly connected architectural units: crystal and cloud. They symbolise the known and the unknown, the clarity of the familiar environment of today and the uncertainty of tomorrow. The entrance is from the town side in the north, the crystal rising towards the built

environment and calling out to it. Raised on 8-metre stilts, the cloud seems to float above the Confluent gardens on the south side – a soft space of hidden streams and countless transitions. It is clad in a metallic envelope that reflects colours and light and captures the echoes of sky and city, water and greenery.

The spatial arrangement of the museum is designed to stimulate the public's curiosity about the present and the future, the known and the still unknown. Ramps and

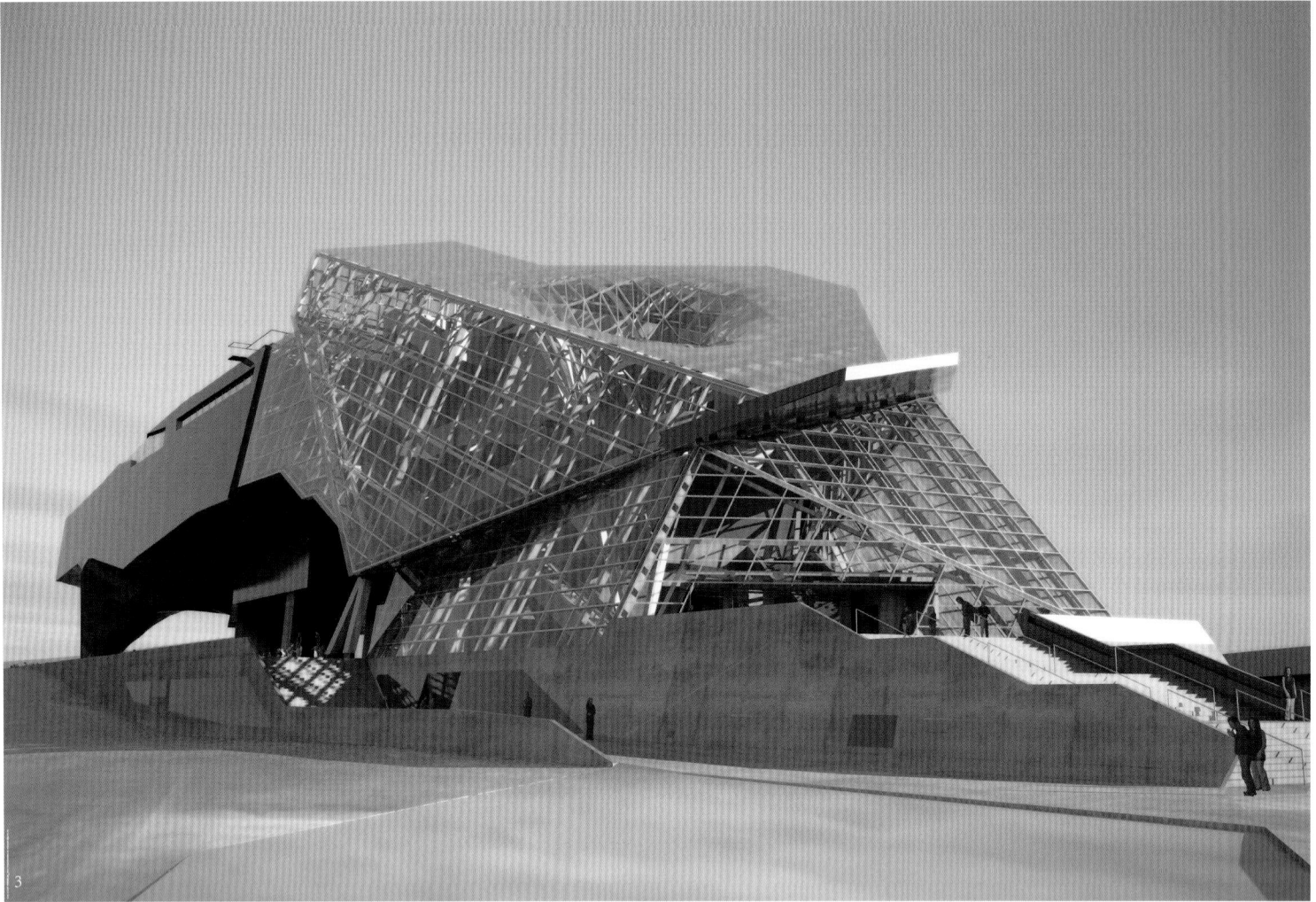

surfaces merge the inside and the outside, resulting in a dynamic sequence of spatial events. This movement continues inside. Closed black boxes and free exhibition areas alternate, making use of the double room height over two levels. The architecture is as changeable as the exhibitions it holds, creating an urban event that is perpetually reinventing itself.

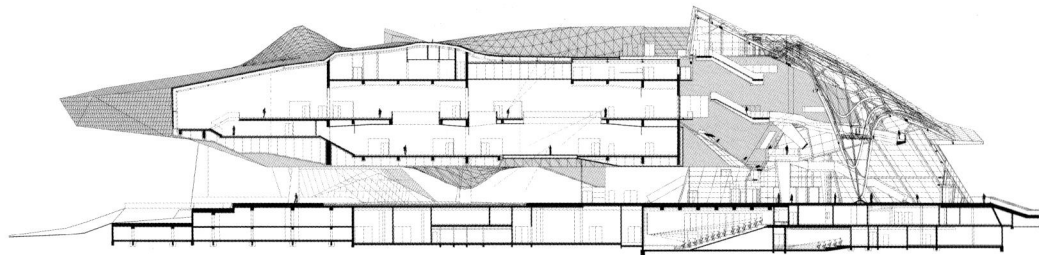

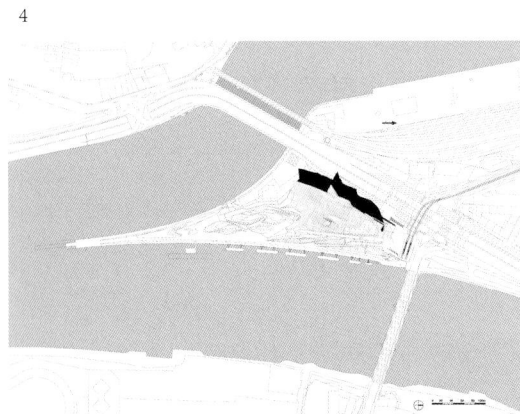

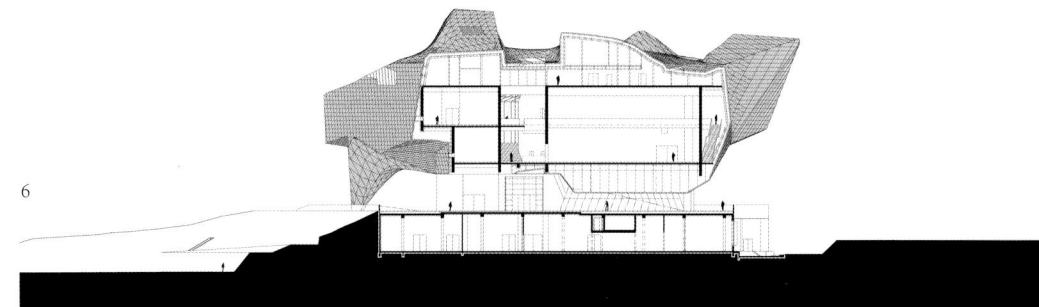

7

1. Confluences Museum – the Crystal Cloud of Knowledge
2. Underneath the Cloud
3. Front view: the Crystal
4. Master plan
5. Longitudinal section
6. Cross section
7. Main entrance
8. Building elements

1. 汇流博物馆——象征知识的水晶云朵
2. 云之下
3. 前景：水晶
4. 总规划图
5. 纵剖面
6. 横截面
7. 主入口
8. 建筑元素

汇流博物馆坐落在里昂半岛的南端，索恩河与罗纳河的汇合处，是一家社科博物馆。未来的问题由技术、生物和道德的变迁领域决定，而这些正是博物馆的中心主题。

项目以直接和互动的方式传递着我们时代的知识，不仅是一座博物馆，还是城市的宝藏。建筑实用而优雅，汇合了博物馆和城市休闲场所，寓教于乐，并且与城市和文化紧密相连。博物馆受到了场地的启发，在河流和城市之间提供了接口，由两个复杂相连的结构组成：水晶和云。它们分别象征着已知与未知，当今熟识环境的清晰感和未来的不确定感。博物馆入口设在北面，朝向城市，水晶楼向着

建成环境上升，发出召唤。坐落在8米高底架上的云楼看似飘浮在南侧的汇流花园（花园到处是暗流和过渡）上一般。它采用金属包层，反射天空、城市、水、绿植的色彩和光线。

博物馆空间结构的设计刺激了公众对现在和未来、已知和未知的好奇心。坡道和平面融合了室内外，形成了动感十足的空间序列。这种运动在内部得到了延续。密闭的黑盒子空间和开放的展览空间相互交替，充分利用了两层楼的高度。建筑可以根据展览内容的不同而改变，形成了一个不断变化的城市景点。

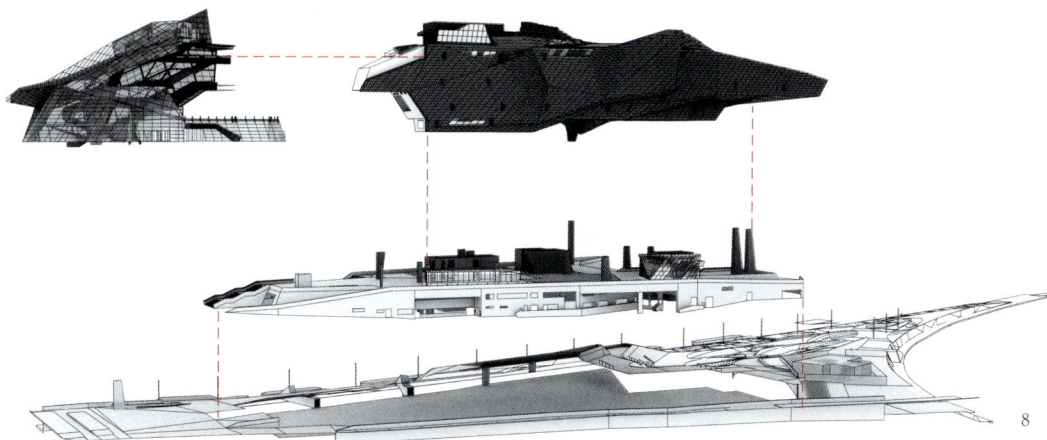

8

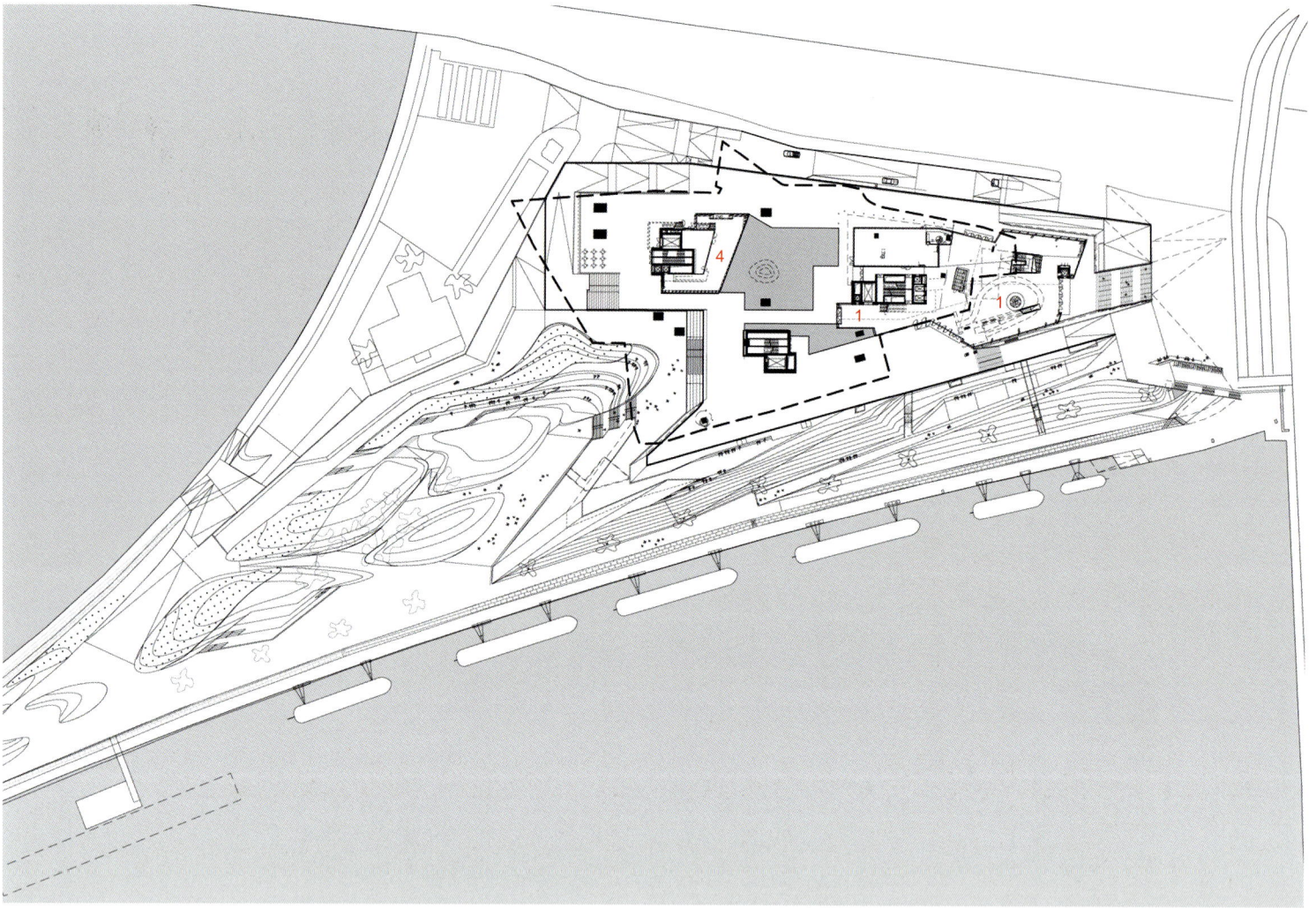

9

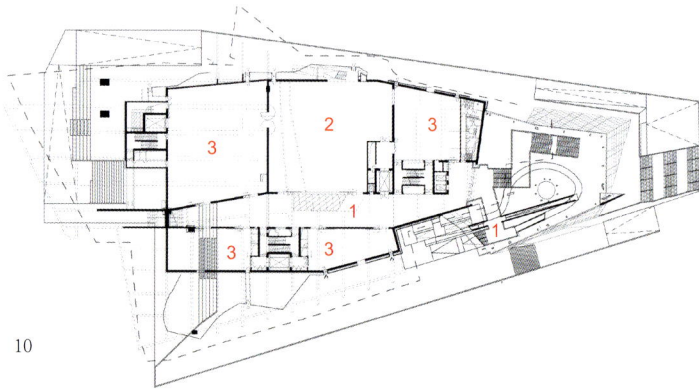

10

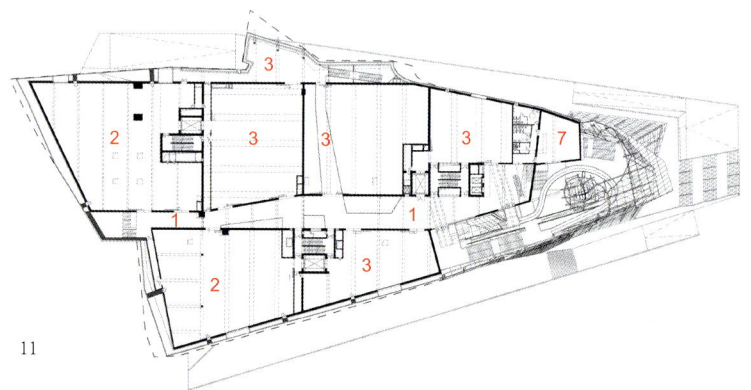

11

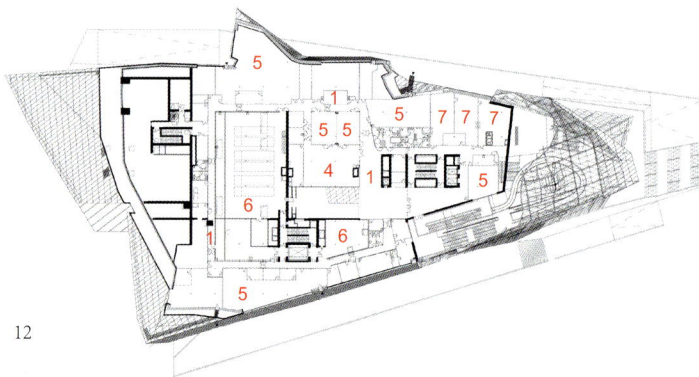

12

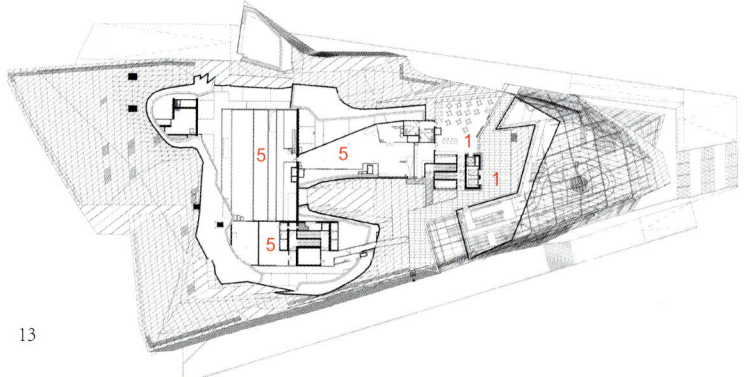

13

French Museum Architecture   316

14

9. Floor plan: Level 00
10. Floor plan: Level +8.84
11. Floor plan: Level +15.64
12. Floor plan: Level +22.78
13. Floor plan: Level +27.03
14. The Crystal: Entrance Hall and Gravity Well
15. Underneath the Cloud

9. 00层平面图
10. +8.84层平面图
11. +15.64层平面图
12. +22.78层平面图
13. +27.03层平面图
14. 水晶里的空间：入口大堂和重力井
15. 云之下

1. Circulation/reception hall
2. Permanent exhibition
3. Temporary exhibition
4. Restaurant
5. Administration office
6. Technical area
7. Workshop

1. 流线区域 / 前台大堂
2. 常设展区
3. 临时展区
4. 餐厅
5. 行政办公室
6. 技术区
7. 工作室

15

REGIONAL FUND
OF CONTEMPORARY ART PACA

PACA当代艺术地域基金会

建筑师 Architect: Kengo Kuma, Toury Vallet
地点 Location: Marseille
完工日期 Completion Date: 2012
图面资料 Visual Documents: Kengo Kuma, Toury Vallet

The project for the Regional Fund of Contemporary Art (FRAC) for the Provence-Alpes-Côte d'Azur region (PACA) is the 3D version of the "museum without walls" invented by the French writer and politician André Malraux. It is a moving and living museum in which the works are in constant flux in order to give the public the greatest chance to see and interact with them. The FRAC was designed as a landmark in the city to give a greater visibility to contemporary art. The building is a symbol with an original and clear identity. It is composed of two clearly defined parts whose shapes fit in to the complex geometry of the plot.

The main body, along Rue Vincent Leblanc, contains the exhibition spaces and documentation centre. A small tower with the auditorium and children's workshop offers an upper terrace on the main boulevard. These two entities are connected by a set of footbridges and are unified by the envelope

of a glass skin, composed of panels with changing opacity.

The project explores the theme of windows and openings on different scales, to give a feeling of welcome and a permanent link with the exterior. The architects wished to produce a living and creative space whose action and effect is felt on the entire city, as well as the surrounding district and neighbourhood. The building adheres to the High Environmental Quality (HQE) charter.

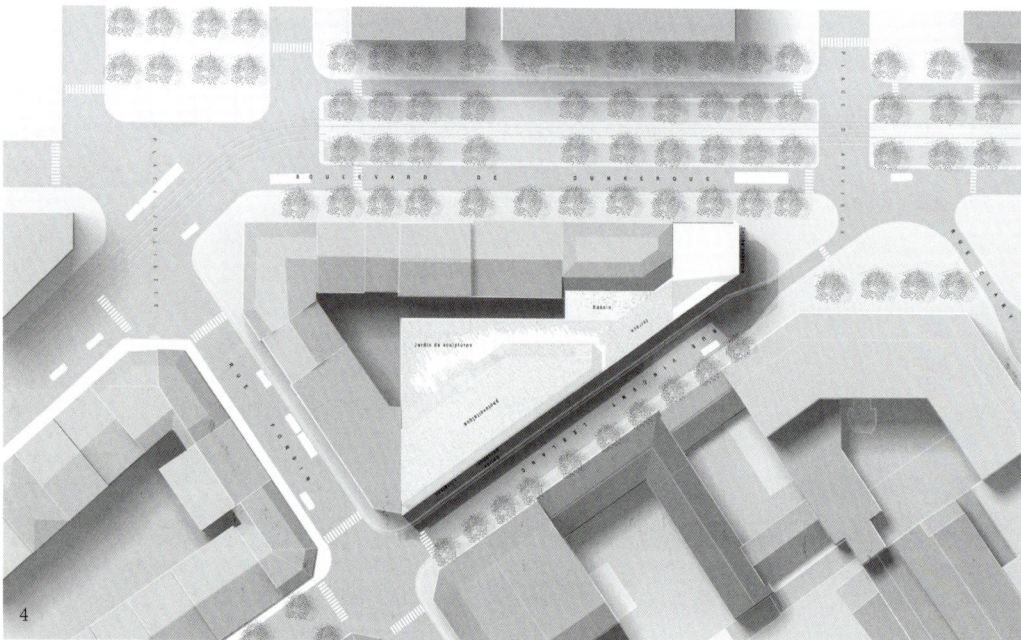

1. Exterior view
2. View of the entrance from the first exhibition room (large module): to the left of the staircase, the pool; to the right the reception desk; at the back, the staircase leading to the second exhibition room (proposals module)
3. View from Rue Vincent Leblanc
4. Master plan

1. 外部图
2. 第一展览室的入口处：楼梯左侧为水池；右侧是前台，后方是通往第二展览室的楼梯
3. 临文森特勒布朗街一景
4. 总规划图

5. Bird's eye view of the urban terrace from Arvieux Square
6. Prow of the edifice: study of the inclination of glass panels on the façade (pivoting on a horizontal axis of pixels)

5. 从阿尔维约广场鸟瞰建筑物前端的城市阳台
6. 大楼前端处理：立面上玻璃板倾斜的研究（沿着水平轴旋转）

普罗旺斯-阿尔卑斯-蓝色海岸（PACA）当代艺术地域基金会（FRAC）是现实版的"无墙博物馆"（由法国作家和政治家安德烈·马尔罗所虚构）。它是一个移动的鲜活博物馆，立面所展出的作品不断变化，以便让公众有更多的机会与艺术品进行互动。当代艺术地域基金会被设计成城市的地标，让当代艺术变得更加普及。建筑拥有独创的鲜明形象，由两个界限分明的部分组成，它们的造型与场地的复杂地形相称。

沿着文森特勒布朗街的博物馆主体内设置着展览空间和档案中心。另一座较小的塔楼内设有礼堂和儿童工坊，在林荫大道上提供了一个上层露台。这两个结构通过一系列人行桥相连，设在同一个玻璃外壳（由透明度变化的面板组成）内。

项目研究了各个尺寸的门窗开口，营造出热情的欢迎氛围并且建立其与外部的永久性联系。建筑师希望打造一个鲜活而创新的空间，让整个城市、周边街区和环境感受到它的活力和影响。建筑完全符合高环境质量标准。

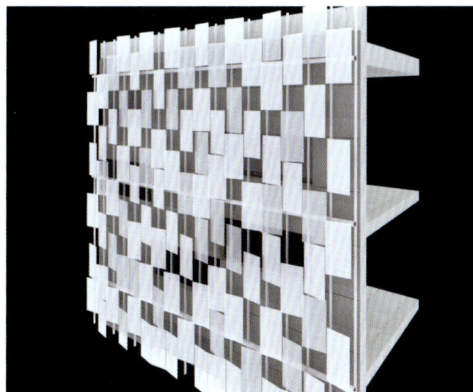

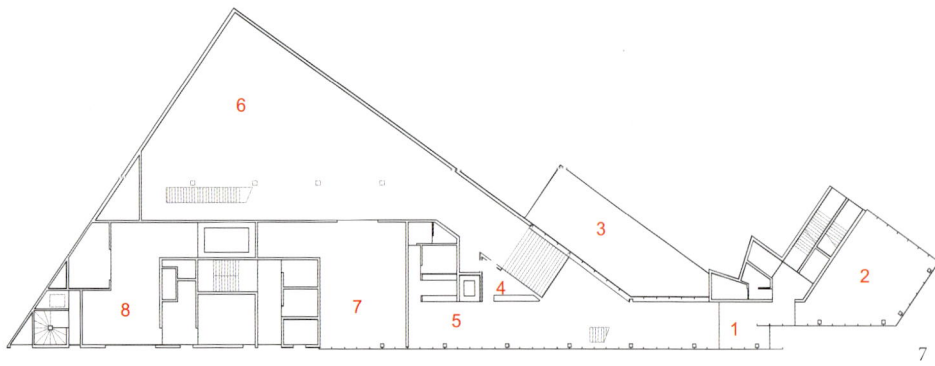

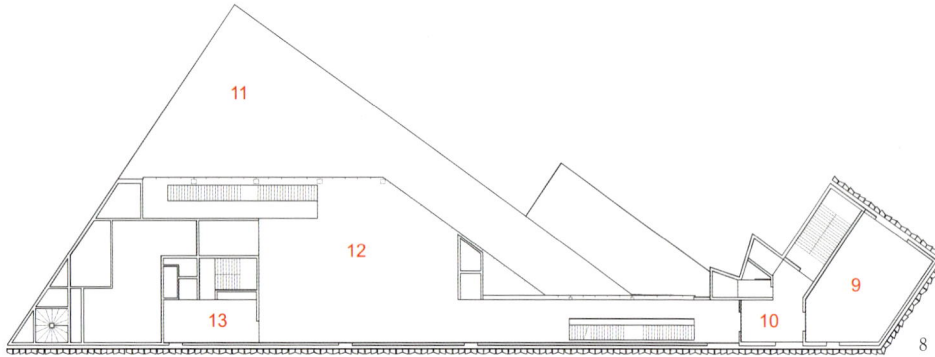

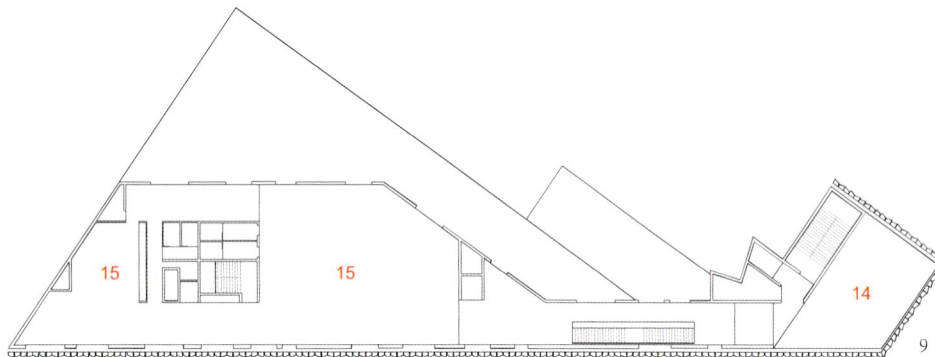

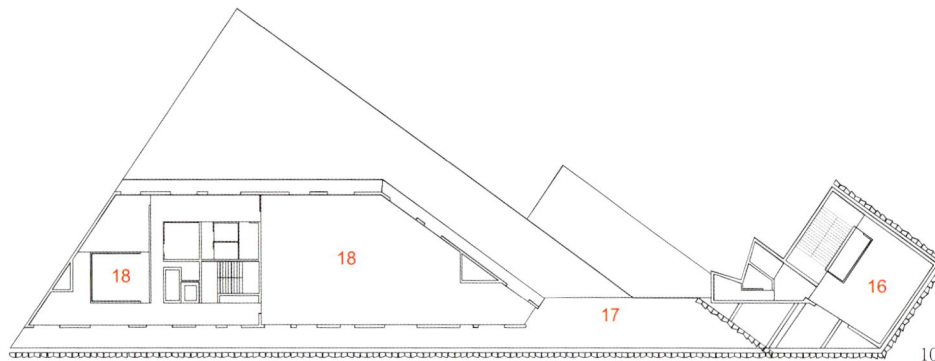

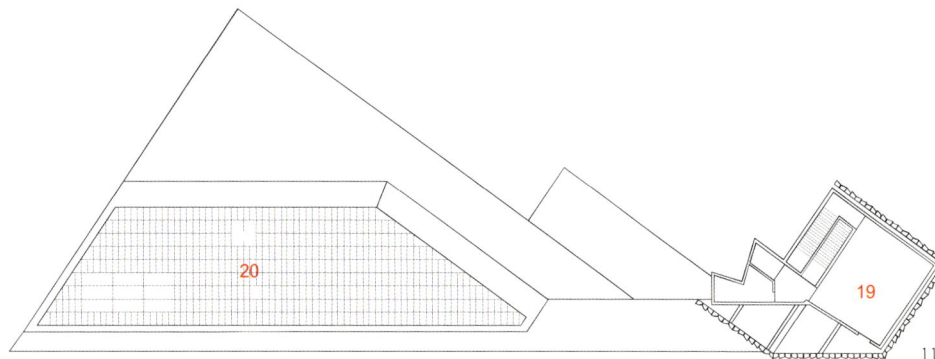

1. Entrance
2. Café
3. Pool
4. Reception
5. Shop
6. Exhibition room 1
7. Workshop
8. Deliveries
9. Conference room
10. Foyer
11. Sculpture garden
12. Exhibition room 2
13. Mounting
14. Urban terrace
15. Resource centre
16. Educational workshop
17. Terrace
18. Offices
19. Exhibition room 3
20. Roof with photovoltaic panels

1. 入口
2. 咖啡厅
3. 水池
4. 前台
5. 商店
6. 展室 1
7. 工作室
8. 送货区
9. 会议室
10. 休息室
11. 雕塑花园
12. 展室 2
13. 安装
14. 城市阳台
15. 资源中心
16. 教学工作室
17. 平台
18. 办公室
19. 展室 3
20. 装置光伏电池板的屋顶

12

13

14

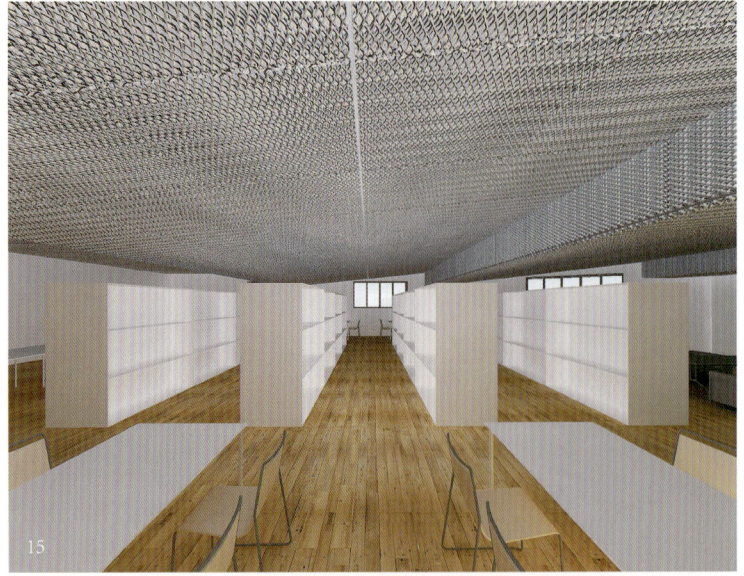

15

16

17

14. Proposals module
15-16. Resource centre
17. Resource centre reception
18. Longitudinal section
19. Arrival at the proposals module

14. 建议模矩空间
15、16. 资源中心
17. 资源中心前台
18. 纵剖面
19. 建议模矩空间的入口

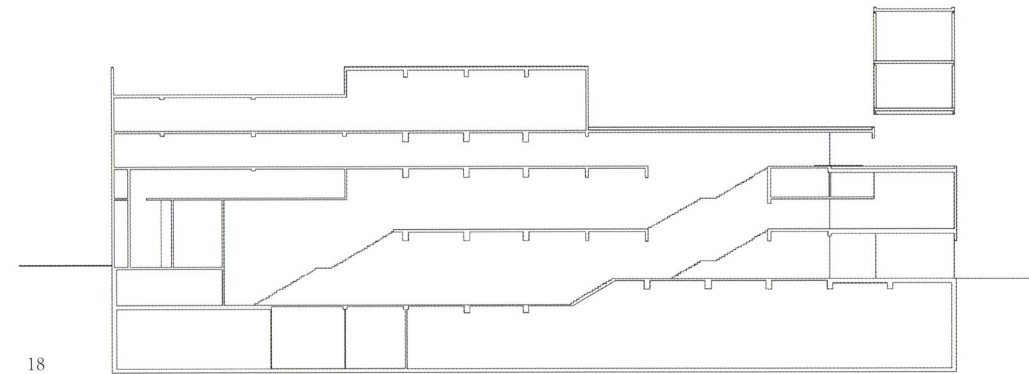

18

MARSEILLE HISTORY MUSEUM

马赛历史博物馆

建筑师 Architect: Roland Carta,
with Studio Adeline Rispal (Exhibition architects)
地点 Location: Marseille
完工日期 Completion Date: 2013
图面资料 Visual Documents: Roland Carta
Architectes & Associés, Studio Adeline Rispal

In the heart of France's oldest city, the Marseille History Museum traces the history of the Phocean city from its foundation, in 600 B.C., to the present day. After the restoration of the adjoining Garden of Archaeological Remains and the establishment of this as the site of the Ancient Port, the museum itself is preparing for change as part of Marseille European Capital of Culture 2013. Around the central theme of seafaring, which has given the daily life of the city its rhythm over the centuries, the new museographic itinerary is divided into 13 historical sequences.

The museum building suffered from a lack of depth and of ceiling height, and there was a dichotomy between the scientific programme and the display area available: "Too many objects, not enough space." It was necessary to generate breathing spaces, vary the ceiling heights and create surprises to strengthen the strong museum sequences. This was done through bringing forward the façade of the

Mediterranean Centre for International Business (CMCI) wing to provide a double-height circulation area; enclosing the west walkway that was previously on the outside of the building; placing the archives in the triple height area under the glass roof; and expanding the garden-level space between the current entrance and the carpark fire escape.

The museum furniture is made up of modular elements called "bales" piled up like merchandise on the port. These

1. Main entrance on Rue Henri Barbusse
2. Start of the walkway
3. The garden of archeological remains
4. Entrance to the museum from the archeological site
5. Architectural treatment of the walkway

1. 临亨利巴比塞街的主入口
2. 闲游行道入口
3. 考古遗迹花园
4. 从考古遗迹进入博物馆的入口
5. 闲游行道的建筑化处理

elements on a human scale will allow life to settle in to the museum, just like the port in the shadow of the great ships. With order and disorder, various time schemes, that of the sea and that of the land, ships and men, the museum expresses all the facets of Marseille life over the course of its history. Tactile and interactive multimedia screens, children's itineraries, resting places to contemplate what one has seen, watch the films that are projected, experience, dream... this furniture can be adapted to all the visitors' needs over time.

坐落在法国最古老城市的中心的马赛历史博物馆追溯了弗坎古城自公元前600年至今的历史。在对旁边的考古遗址园进行修复并将场地建立为古港口之后，博物馆准备进行改造，以成为2013马赛欧洲文化之都活动的一部分。航海是马赛上千年来一直不变的生活主题，博物馆以此为主题，将展览划分成13个历史序列。

博物馆的建筑缺乏深度和棚顶高度，并且在科学项目和可用展览区上有所冲突：太多藏品，空间不够。有必要建造一些休息空间，变化棚顶高度并创造惊喜来增强博物馆展览序列的存在感。建筑师将国际商务地中海中心（CMCI）前移，打造了一个两层楼高的流通空间；闭合了建筑外面西侧的走道；将档案馆设在玻璃屋顶下一个三层楼高的区域；并且扩展了现有入口和停车场防火通道之间的花园空间。

博物馆内的装置全部由名为"大包"的模块化元素组成，就像港口上的货物一样。这些具有人性尺度的元素为博物馆注入了活力，使其好像是船只往来的港口。整齐和混乱、错杂的时间点、海陆两方面、船舶和人类……博物馆展现了历史上马赛生活的方方面面。触摸互动多媒体屏幕、儿童参观活动、休息区、投影影片、体验、梦想……博物馆装置可以根据参观者的需求而进行改造。

6. Entrance to the museum from the business centre
7. Display cases punctuate the space and direct visitors towards the ancient port
8. Reception area
9. The architectural display cases are lined up like boats in port

6. 从商务中心进入博物馆
7. 展示柜装点着展览空间，也指引游客走向古老的港口
8. 接待区
9. 与建筑结构融合的展示柜有如港口的船舶——并排着

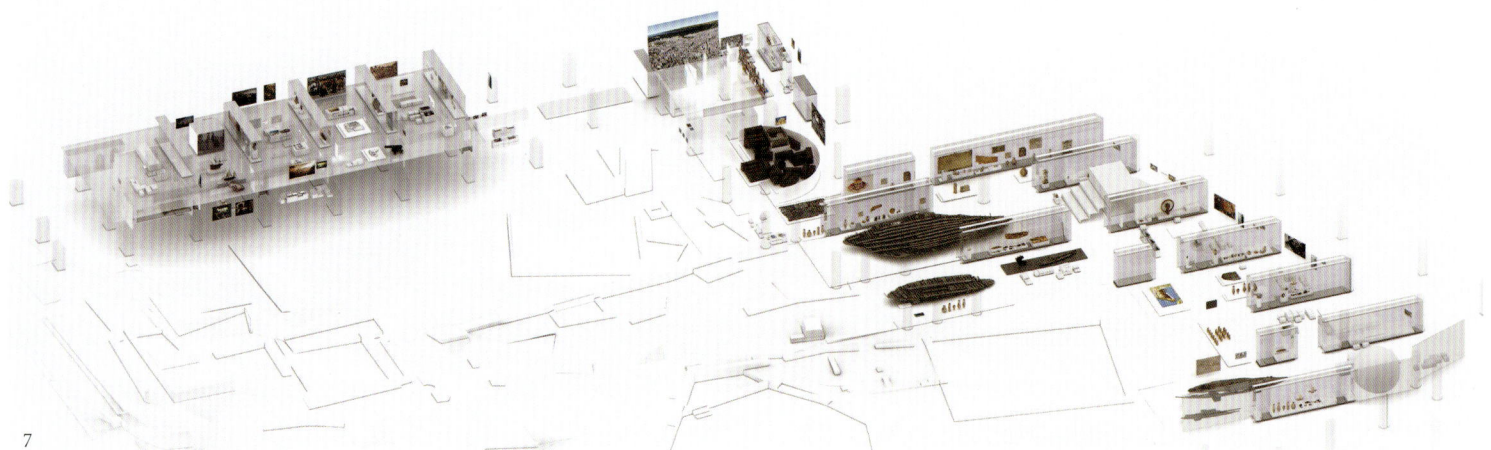

10. The ancient wrecks are displayed as if they are in a naval shipyard
11. A double-height space links the modern and contemporary galleries
12. The new façade frames views of the city
13. Spatial organisation of the permanent exhibition rooms

10. 古代沉船陈列，仿佛在海军造船厂
11. 一个双层高的空间连接了现代和当代画廊
12. 新建筑立面将城市景色框在其中
13. 常设展览室的空间组织

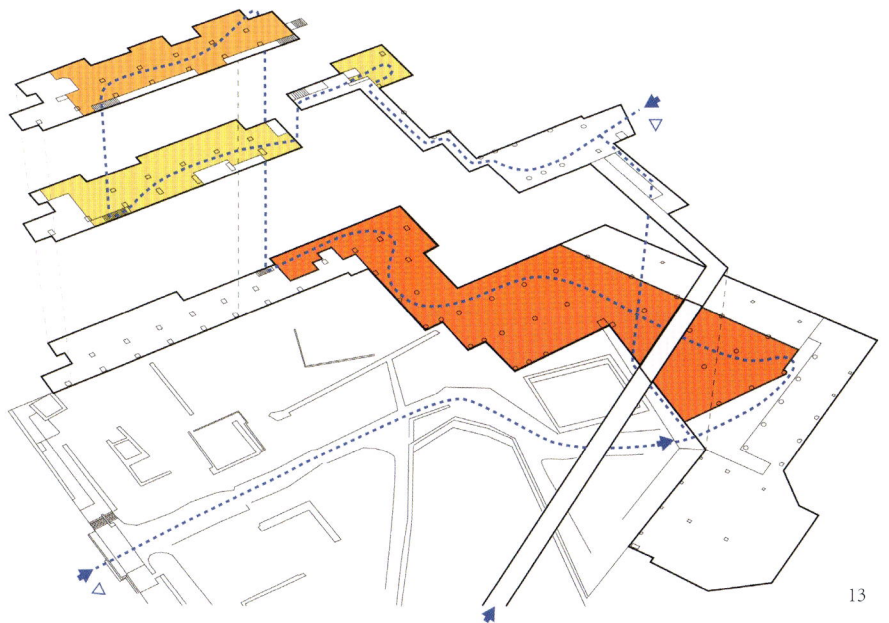

MUSEUM OF THE CIVILISATIONS OF EUROPE AND THE MEDITERRANEAN

欧洲与地中海文明博物馆

建筑师 Architect: Rudy Ricciotti, Roland Carta, Studio Adeline Rispal (Exhibition architects)
地点 Location: Marseille
完工日期 Completion Date: 2013
图面资料与照片 Visual Documents: Agence Rudy Ricciotti, Lisa Ricciotti

The Museum of the Civilisations of Europe and the Mediterranean (MuCEM) will be launched in 2013 as part of Marseille's year as European Capital of Culture. A major project of 40,000m², it is spread over three sites at the entrance to the Old Port. The heart of the museum will be the new 15,000m² building, constructed on the old J4 jetty by the architect Rudy Ricciotti and linked to the Saint-Jean fort by a footbridge. The foundation of the project is an urban vision that carves out its identity in strong lines. The ground floor with its sociological functions will face the sea esplanade and the Mediterranean wholesale market, and the volumetry will be horizontal so as not to rival the Saint-Jean fort.

Views, sea, sun and stone had to be orchestrated in a unifying and well-conceived programme for this museum.

First of all, a perfect 72m square. In this square, another, 52m square, is drawn, to contain the exhibition and conference rooms that are the heart of the museum. Around this, above and below, are the service areas. But the empty spaces between the two all look to the central square and form connections, following the paths of the old lanes of the town. Lured by the view of the fort, the sea or the port, most visitors will choose this

1. Perspective drawing – view from the sea
2. Façade
3. Perspective drawing – footbridge
4. Perspective drawing – view of the interior
5. Master plan
6. Perspective drawing – between two exterior footbridges

1. 博物馆外观模拟效果图
2. 博物馆立面
3. 人行桥透视图
4. 内景透视图
5. 总规划图
6. 两座外部人行桥之间的空间透视图

route to arrive at the museum. Taking one of the two interlaced ramps, they will then plunge into an imaginary Tower of Babel or ziggurat in order to climb up to the roof and, in the mind's eye, all the way up to the Saint-Jean fort. This peripheral fault offers a breathing space that takes one momentarily away from the museum, in the salty scent from the nearby seawater ditches, in order to chase away any doubts one might have about the way the history of our civilisations has been used.

The tectonic choice of a high quality concrete straight from the latest French industrial research makes its mineral mark on the high ramparts of the Saint-Jean fort. A single material with the dun colour of dust softened by the golden light, it is a eulogy for the dense and fragile environment around it. The MuCEM will seem to rise up from a landscape of stone, and look east through its shadows. But on the harbour basin side at the height of the pontoon, a bright slanting light casting silver reflections on the blue water will penetrate this territory so close to the sea.

作为马赛欧洲文化之都活动的一部分，欧洲与地中海文明博物馆将于2013年正式对外开放。项目总面积40,000平方米，在旧港口入口处绵延三个场地。博物馆中心的新建筑由鲁迪·里西奥第操刀设计，总面积15,000平方米，建在J4码头上，与圣让堡垒通过人行桥相连。项目的设计具有城市视野，以强烈的线条来彰显自己的身份。一楼的社会功能区面朝海滨大道和地中海批发市场。水平向的建筑结构不会让圣让港口失色。

风景、海洋、日光和石头在精心设计的博物馆项目中显得和谐而统一。建筑师首先画出了一个长72米的广场。广场内部还有一个长52米的正方形，用以设置展览厅和会议室。环绕着中央结构的是服务区。二者之间的空敞空间都朝向中央的广场并相互连接，遵循了古城的道路。受到港口、海洋或港口景色的吸引，大多数参观者会选择这条路线进入博物馆。走过交错的坡道，他们将登上虚幻的巴比伦塔，到达屋顶，将圣让港口的美景尽收眼底。这种边缘断层结构所提供的优美景色让人们暂时脱离了博物馆，在咸咸的海风中追逐我们文明的历史。

在构造上，建筑师选用的高品质混凝土直接来自于最新的法国工业研究，使其成为了圣让港口的标志性建筑。单一的材料和灰暗的色彩中点缀了一丝金光，让建筑反映了周边密集而脆弱的环境。欧洲与地中海文明博物馆看似崛起于一片岩石景观之中，在东面洒下了阴影。但是在港池一侧，与浮桥等高，一道明亮的光线在蔚蓝的海水上投下了银色的倒影，让博物馆与大海融为一体。

7. The museum under construction, view from the sea – May 2012
8. Exterior view – latticework panel
9. Construction details

7. 2012年5月建造中的博物馆
8. 格栅组构成的建筑外观
9. 建造细部

Index of Museums 博物馆资讯

Matisse Museum
Musée Matisse
Palais Fénelon
59360 Le Cateau-Cambrésis
T +33 (0)3 59 73 38 00
F +33 (0)3 59 73 38 01
http://museematisse.cg59.fr/
museematisse@cg59.fr

Meaux Country Museum of the Great War
Musée de la Grande Guerre du Pays de Meaux
Rue Lazare Ponticelli 77100 Meaux
T +33 (0)1 60 32 14 18
www.museedelagrandeguerre.eu

Mobile Art Pavilion
Pavillon Mobile Art
Institute of the Arab World (Institut du Monde Arabe)
1 rue des Fossés Saint-Bernard
Place Mohammed V
75236 Paris Cedex 05
T +33 (0)1 40 51 38 38
F +33 (0)1 43 54 76 45
www.imarabe.org

**Museum of the Civilisations of Europe
and the Mediterranean**
Musée des Civilisations d'Europe
et de Méditerranée (MuCEM)
Caserne du Muy
21 rue Bugeaud 13003 Marseille
www.mucem.org

Nancy Museum of Fine Arts
Musée des Beaux-Arts de Nancy
3 place Stanislas 54000 Nancy
T +33 (0)3 83 85 30 72
F +33 (0)3 83 85 30 76
mbanancy@mairie-nancy.fr
http://mban.nancy.fr

Orangerie Museum
Musée de l'Orangerie
Jardin des Tuileries 75001 Paris
T +33 (0)1 44 77 80 07
information.orangerie@musee-orangerie.fr
www.musee-orangerie.fr

Orsay Museum
Musée d'Orsay
62 rue de Lille 75007 Paris
T +33 (0)1 40 49 48 14
www.musee-orsay.fr

Pompidou Centre
Centre Pompidou
Place Georges Pompidou 75004 Paris
T +33 (0)1 44 78 12 33
www.centrepompidou.fr

President Jacques Chirac Museum
Musée du Président Jacques Chirac
19800 Sarran
T +33 (0)5 55 21 77 77
F +33 (0)5 55 21 77 78
musee.president@cg19.fr
www.museepresidentjchirac.fr

Quai Branly Museum
Musée du Quai Branly
37 quai Branly 75007 Paris
T +33 (0)1 56 61 70 00
communication@quaibranly.fr
www.quaibranly.fr

Quimper Museum of Fine Arts
Musée des Beaux-Arts de Quimper
40 place Saint-Corentin 29000 Quimper
T +33 (0)2 98 95 45 20
F +33 (0)2 98 95 87 50
musee@mairie-quimper.fr
www.mbaq.fr

Regional Fund of Contemporary Art PACA
Fond Régional d'Art Contemporain PACA
1 place Francis Chirat 13002 Marseille
T +33 (0)4 91 91 27 55
F +33 (0)4 91 90 28 50
infos@fracpaca.org
www.fracpaca.org

**Strasbourg Museum of Modern and
Contemporary Art**
Musée d'Art Moderne et Contemporain
de Strasbourg
1 place Hans Jean Arp 67000 Strasbourg
T +33 (0)3 88 23 31 31
www.musees.strasbourg.eu

Tomi Ungerer Museum
Musée Tomi Ungerer
Villa Greiner
2 avenue de la Marseillaise
67076 Strasbourg cedex
T +33 (0)3 69 06 37 27
F +33 (0)3 69 06 37 28
www.musees.strasbourg.eu

Val-de-Marne Museum of Contemporary Art
Musée d'Art Contemporain du Val-de-Marne
(Mac-Val)
Place de la Libération BP 147
94404 Vitry-sur-Seine Cedex
T +33 (0)1 43 91 64 20
F +33 (0)1 43 91 64 30
contact@macval.fr
www.macval.fr

Valence Museum of Fine Arts and Archeology
Musée des Beaux-Arts et Archéologie
4 place des Ormeaux 26000 Valence
T +33 (0)4 75 79 20 80
F +33 (0)4 75 79 20 84
info@musee-valence.org
www.musee-valence.org

Vesunna Gallo-Roman Site-Museum
Vesunna, Site-Musée Gallo-Romain
Parc de Vésone
20 rue du 26e Régiment d'Infanterie
24000 Périgueux
T +33 (0)5 53 53 00 92
F +33 (0)5 53 35 40 12
vesunna@perigueux.fr
www.perigueux.fr

Würth Museum
Musée Würth
Z.I. Ouest
Rue Georges Besse BP 40013
67158 Erstein Cedex
T +33 (0)3 88 64 74 84
F +33 (0)3 88 64 74 88
www.musee-wurth.fr
mwfe.info@wurth.fr

Index of Architects 建筑师资讯

Amplitude Architectes
14 rue Génissieu 38000 Grenoble
T +33 (0)4 76 12 25 84
F +33 (0)4 76 12 90 36
contact@amplitude-architectes.com
www.amplitude-architectes.com

Architecture Studio
10 rue Lacuée 75012 Paris
T +33 (0)1 43 45 18 00
F +33 (0)1 43 43 81 43
as@architecture-studio.fr
www.architecture-studio.fr

Ateliers 234
234 rue du Faubourg Saint-Antoine 75012 Paris
T +33 (0)1 55 25 15 10
234@fbg234.com
www.a234.fr

Ateliers AFA – Adrien Fainsilber & Associés
69 rue Barrault 75013 Paris
T +33 (0)1 45 65 50 90
F +33 (0)1 45 65 50 91
agence@ateliers-afa.eu
www.ateliers-afa.eu

Atelier de l'Ile
3 rue Dagorno 75012 Paris
T +33 (0)1 48 06 22 00
F +33 (0)1 48 06 91 75
paris.atile@atile.fr
www.atile.fr

Beaudouin-Husson Architectes
3 rue de la Monnaie 54000 Nancy
T +33 (0)3 83 36 49 46
F +33 (0)3 83 36 80 30
elb@beaudouin-architectes.com
www.beaudouin-architectes.fr

Bodin & Associés
33 rue des Francs-Bourgeois 75004 Paris
T +33 (0)1 44 59 21 40
F +33 (0)1 44 59 21 45
33@bodin.fr
www.bodin.fr

Brochet Lajus Pueyo
Hangar G2 - Quai Armand Lalande 33300 Bordeaux
T +33 (0)5 57 19 59 19
F +33 (0)5 57 19 59 10
architectes@brochet-lajus-pueyo.fr
www.brochet-lajus-pueyo.fr

Riland Carta Architectes & Associés
20 rue Saint Jacques 13006 Marseille
T +33 (0)4 96 10 29 00
F +33 (0)4 96 10 29 09
agence@cplust.eu
www.cplust.eu

Clément-Vergely Architectes
7 quai Général Sarrail 69006 Lyon
T +33 (0)4 72 65 91 44
F +33 (0)4 72 65 92 47
atelier@vergelyarchitectes.com
www.vergelyarchitectes.com

Coop Himmelb(l)au
Wolf D. Prix & Partner ZT GmbH
Spengergasse 37
1050 Vienna, Austria
T +43 (0)1 546 60
F +43 (0)1 546 60-600
communications@coop-himmelblau.at
www.coop-himmelblau.at

ECDM – Emmanuel Combarel Dominique Marrec
7 passage Turquetil 75011 Paris
T +33 (0)1 44 93 20 60
F +33 (0)9 58 72 33 21
contact@ecdm.fr
http://ecdm.eu/

Ferrero & Rossi
4 avenue de la Résistance 06140 Vence
T +33 (0)4 93 58 15 03
F +33 (0)4 93 58 21 38
contact@ferrero-architecte.com
contact@rossi-architecte.com
www.ferrero-architecte.com

Jacques Ferrier architectures
77 rue Pascal 75013 Paris
T +33 (0)1 43 13 20 20
F +33 (0)1 43 13 20 21
communication.ferrier@agencejfa.com
www.jacques-ferrier.com

Manuelle Gautrand
36 boulevard de la Bastille 75012 Paris
T +33 (0)1 56 95 06 46
F +33 (0)1 56 95 06 47
contact@manuelle-gautrand.com
www.manuelle-gautrand.com

Zaha Hadid Architects
10 Bowling Green Lane
London EC1R 0BQ United Kingdom
T +44 20 7253 5147
F +44 20 7251 8322
press@zaha-hadid.com
www.zaha-hadid.com

Steven Holl Architects
450 West 31st street 11th floor
New York, NY 10001, USA
T +1 212 629 7262
F +1 212 629 7312
nyc@stevenholl.com
www.stevenholl.com

Frédéric Jung
20 rue Frédérick Lemaître 75020 Paris
T +33 (0)1 42 02 04 84
F +33 (0)1 42 02 03 10
jung.architecture@free.fr

Kuma & Associates Europe
16 rue Martel 75010 Paris
T +33 (0)1 44 88 94 90
F +33 (0)1 42 46 23 55
kuma@kkaa.co.jp
http://kkaa.co.jp

Christophe Lab (Atelier)
21 rue de Tanger 75019 Paris
T +33 (0)1 53 26 56 57
57 Familistère aile gauche 02120 Guise
T+33 (0)3 23 09 25 39
atelierlab@wanadoo.fr

Moatti-Rivière
22 rue de Paradis 75010 Paris
T +33 (0)1 45 65 44 04
communication@moattiriviere.com
www.moatti-riviere.com

Moget-Gaubert Architectes
7 rue Suger 75006 Paris
T +33 (0)6 21 33 52 56
T +33 (0)6 16 79 73 75
pierre.moget@free.fr

Jean Nouvel (Ateliers)
10 cité d'Angoulême 75011 Paris
T +33 (0)1 49 23 83 83
F +33 (0)1 43 14 81 15
info@jeannouvel.fr
www.jeannouvel.com

Renzo Piano Building Workshop
34 rue des Archives 75004 Paris
T +33 (0)1 44 61 49 00
F +33 (0)1 42 78 01 98
france@rpbw.com
www.rpbw.com

Jean-Paul Philippon
6 rue de Braque 75003 Paris
T +33 (0)1 42 71 02 96
F +33 (0)1 42 71 94 43
j.p.philippon@wanadoo.fr
www.philippon-architecte.fr

Projectiles
8 Passage Brûlon 75012 Paris
T +33 (0)1 58 30 82 61
F +33 (0)1 43 67 86 47
atelier@projectiles.net
www.projectiles.net

Jacques Ripault Architecture
43 rue des Tournelles 75003 Paris
T +33 (0)1 44 54 29 30
F +33 (0)1 44 54 29 39
architectes@ripaultarchitecture.fr
www.jacquesripault.com

Rudy Ricciotti
17 boulevard Victor Hugo 83150 Bandol
T +33 (0)4 94 29 52 61
F +33 (0)4 94 32 45 25
rudy.ricciotti@wanadoo.fr
www.rudyricciotti.com

Adeline Rispal (Studio)
23 rue du Faubourg Saint-Denis 75010 Paris
T +33 (0)1 43 56 91 45
contact@adelinerispal.com
www.adelinerispal.com

Rogers Stirk Harbour + Partners
Thames Wharf, Rainville Road
London W6 9HA United Kingdom
T +44 (0)20 7385 1235
F +44 (0)20 7385 8409
enquiries@rsh-p.com
www.rsh-p.com

Toury Vallet Architectes
14 rue Lauzin 75019 Paris
T+33 (0)1 44 75 37 69
F+33 (0)1 44 75 38 40
mail@touryvallet.fr
www.touryvallet.f r

Wilmotte & Associés SA
68 rue du Faubourg Saint-Antoine 75012 Paris
T+33 (0)1 53 02 22 22
wilmotte@wilmotte.fr
www.wilmotte.com

图书在版编目（CIP）数据

法国博物馆建筑 / 法国亦西文化编著；常文心译. --
沈阳：辽宁科学技术出版社，2012.11
　　ISBN 978-7-5381-7714-5

　　Ⅰ. ①法… Ⅱ. ①法… ②常… Ⅲ. ①博物馆－建
筑设计－法国 Ⅳ. ①TU242.5

　　中国版本图书馆CIP数据核字(2012)第242522号

出版发行：辽宁科学技术出版社
　　　　　（地址：沈阳市和平区十一纬路29号　邮编：110003）
印　刷　者：利丰雅高印刷（深圳）有限公司
经　销　者：各地新华书店
幅面尺寸：230mm×290mm
印　　张：43
插　　页：4
字　　数：50千字
印　　数：1～2000
出版时间：2012年 11 月第 1 版
印刷时间：2012年 11 月第 1 次印刷
责任编辑：陈慈良　隋　　敏
封面设计：杨春玲
版式设计：杨春玲
责任校对：周　　文
书　　号：ISBN 978-7-5381-7714-5
定　　价：328.00元

联系电话：024-23284360
邮购热线：024-23284502
E-mail: lnkjc@126.com
http://www.lnkj.com.cn
本书网址：www.lnkj.cn/uri.sh/7714